Charles Frederick Cross, Edward John Bevan

Cellulose

An Outline of the Chemistry of the Structural Elements of Plants....

Charles Frederick Cross, Edward John Bevan

Cellulose
An Outline of the Chemistry of the Structural Elements of Plants....

ISBN/EAN: 9783337025311

Printed in Europe, USA, Canada, Australia, Japan

Cover: Foto ©berggeist007 / pixelio.de

More available books at **www.hansebooks.com**

CELLULOSE

AN OUTLINE OF THE CHEMISTRY OF

THE STRUCTURAL ELEMENTS OF PLANTS

WITH REFERENCE TO THEIR

NATURAL HISTORY AND INDUSTRIAL USES

BY

CROSS & BEVAN

(C. F. CROSS, E. J. BEVAN, AND C. BEADLE)

NEW IMPRESSION

LONGMANS, GREEN, AND CO.
39 PATERNOSTER ROW, LONDON
NEW YORK, BOMBAY, AND CALCUTTA
1910

PREFACE

TO

THE SECOND EDITION

Some of the published criticisms of the book appearing to indicate a certain misunderstanding of its plan and purpose, we think it opportune to further and more specifically explain.

The chemistry of cellulose is necessarily a chemistry of colloidal, uncrystallisable substance : hence the relationships of any derivative to the parent substance is not established by a comparison of composition and properties, but requires a complete statistical account of the reaction. Even then it is generally impossible to measure attendant changes of molecular condition, or, to put it more precisely, changes of dimensions of the reacting unit, accompanying the formation of derivative compounds.

It is consequently impossible in the present state of our knowledge to write a chemistry of cellulose in the terms of the thoroughly systematised branches of the science. We have therefore adopted the plan of classifying the empirical subject-matter, as a necessary first step in progress from the *purely* empirical. From the methodical treatment of the subject-matter —under such heads as Hydrolysis, Oxidations, Ester formation, and other special synthetic reactions—a certain order has been introduced, from which there result at least prominent suggestions of the underlying constitutional relationships.

So far has this proceeded that it is possible at this date to propose constitutional formulæ for even so complex a body as a lignocellulose, such as would be a consistent summary of the experimental facts as to composition and reactions. It would, however, be premature to attempt this in any other way than as a working hypothesis to guide investigations. There remains, indeed, still unsolved the problem of the actual condition of matter in these complex colloidal forms relatively to the gaseous and liquid states, and pending a solution of this major problem, we can only continue on the basis of progressive empiricism which we originally adopted.

The present is therefore in the main a reprint of the former edition with a small portion of the text rewritten and the addition of an appendix giving an account of more important recent contributions.

4 NEW COURT, LONDON, W.C.
February 25, 1903.

PREFACE

THE purpose of this short work on a large subject is to consolidate the scattered contributions of investigators, with more especial reference to the work of the past fifteen years. By this later work the subject has been considerably widened, not merely through the growth of the subject-matter, but by notable additions of experimental methods on the one hand, and theoretical generalisations on the other. In reviewing the present position of cellulose chemistry the work has taken the form of a monograph as distinguished from a textbook. In adopting the freer form of writing we have reserved a certain latitude of treatment, in view of the fact that 'Cellulose' has not yet been accorded a definite position in the specialised sections of organic chemistry, and also because we find it necessary to address ourselves to original workers, and from time to time to point out with particular emphasis the weaker points in the evidence for the theoretical conclusions. At such points suggestions are given of subject-matter for further research, which it is our desire to stimulate. In presenting this work, on the other hand, to the masters and 'past masters' of the science it would have been out of place to have adopted the more positive method of a textbook, or to have entered into the more minute detail of a handbook.

In the incidental treatment of the technology of the subject we have endeavoured to maintain the scientific perspective rather than to discuss the practical details of processes.

The photo-micrographs included in the work are from the expert handiwork of our friend Mr. J. CHRISTIE, F.R.M.S., of 72 Mark Lane, E.C., to whom we record our best thanks for this interesting addition to the subject-matter.

The book is printed upon a paper carefully selected as composed of the 'normal' celluloses, and to the exclusion of the inferior 'celluloses' ordinarily employed for the manufacture of printing papers. Upon the reasons for this preference we have something to say in the text (p. 305).

We have to thank our friend Mr. J. C. CHORLEY for contributions of experimental results, and for kind assistance in connection with the proofs.

4 NEW COURT, LINCOLN'S INN, LONDON, W.C.

CONTENTS

PART I

THE TYPICAL CELLULOSE AND THE CELLU-
LOSE GROUP 1

PART II
COMPOUND CELLULOSES

LIGNOCELLULOSES 89
PECTOCELLULOSES AND MUCOCELLULOSES . . . 214
ADIPOCELLULOSES AND CUTOCELLULOSES 225

PART III
EXPERIMENTAL AND APPLIED . . 242

APPENDIX I 311
PHOTO-MICROGRAPHS 312
APPENDIX II (1903) 313
INDEX OF AUTHORS 321
INDEX OF SUBJECTS 323

CELLULOSE

PART I

THE TYPICAL CELLULOSE AND THE CELLULOSE GROUP[1]

CELLULOSE is the predominating constituent of plant tissues, and may be shortly described as the structural basis of the vegetable world. The ordinary flowering plant is a complex structure, and its several parts are also complex—that is, are made up of cells. These cells exhibit an infinite variety of form, the main lines of differentiation necessarily conforming with variations in function. The growing cell is, of course, nitrogenous, the living functions depending upon its protoplasmic contents. What we have to deal with, however, is the cell, less the cell-contents of whatever kind, whether 'organic'—that is, concerned in the assimilating or other living functions of the cell—or of the nature of by-products of metabolism excreted or thrown off from the main stream of matter undergoing elaboration into the essential structures of the plant. We have to deal, in fact, with the cell-wall or envelope, to which the term *cellulose* has been applied as to a chemical individual. There are, as might be expected, a great many varieties of cellulose, and the term must be taken as denoting a chemical group. The celluloses, taken as a

[1] Sometimes abbreviated to the short title 'Cellulose.'

Cellulose

group, present the following characteristics :—colourless substances insoluble in all simple solvents, generally but variably resistant to processes of oxidation and hydrolysis, non-nitrogenous, and having the empirical constitution characteristic of the *carbohydrates*, i.e. $C_nH_{2m}O_m$. Their reactions are those of 'saturated' compounds. Their empirical formulæ and relationships to the carbohydrates of low molecular weight further indicate 'single-bond' linking of their C atoms as exclusively prevailing. It must be noted here that the typical celluloses are not separated from the plant in a 'pure' state, but in admixture or in intimate chemical union with other compounds or groups of compounds. The latter are distinguished by greater reactivity, e.g. they readily yield to alkaline hydrolysis ('pectic' bodies), to oxidation (colouring matters), or to the action of the halogens. In the latter is included the very important group of lignified celluloses or lignocelluloses (woods) distinguished by the presence of keto-hexene groups in union with the cellulose, and therefore combining directly with the halogens. These points are sufficient to indicate the principles underlying the general method adopted in the laboratory for the isolation of cellulose from vegetable raw material which consists in (1) Alkaline hydrolysis— boiling the tissue or fibre in solution of alkaline hydrates (1-2 p.ct. NaOH), and after washing (2) exposure to the action of a halogen—chlorine gas or bromine water at the ordinary temperature, and (3) second alkaline hydrolysis—boiling in alkaline solutions, e.g. sodium sulphite, carbonate, or hydrate, to complete the resolution and to dissolve away the products formed from the non-cellulose constituents by the preceding treatment.

In the case of very refractory substances such as the hard woods, it is sometimes necessary to repeat the treatment with the halogen. After such treatment and thorough washing the material is treated exhaustively with alcohol and with ether to remove fatty or resinous by-products of the oxidation.

Cellulose obtained in this way from raw fibrous materials, e.g. cotton, flax, hemp, ramie, is a white substance distinguished by more or less lustre and translucency, retaining the structural characteristics of the raw material, of 1·5 sp.gr., and as a chemical individual distinguished amongst C.H.O compounds by its negative or non-reactive characteristics.

In the brief account which ensues, of the general chemistry of cellulose, cotton-cellulose is taken as the type. The points of differentiation of other members of the group from the type will be noted subsequently.

The empirical composition of the pure cellulose is represented by the percentage numbers

$$
\begin{array}{lr}
C & 44·2 \\
H & 6·3 \\
O & 49·5 \\
\end{array}
$$

corresponding with the statistical formula $C_6H_{10}O_5$. These numbers represent the composition of the 'ash-free' cellulose. All vegetable tissues contain a greater or less proportion of inorganic constituents of which a certain proportion are retained by the cellulose isolated, as described, or by any of the processes practised on the large scale in the arts (*infra*). The celluloses burn with a quiet luminous flame, leaving these inorganic constituents as an ash, retaining more or less the form of the original. In cotton the average proportion of ash is 0·1–0·4 p.ct. The composition of the ash has, no doubt, certain specific relationships to the several celluloses, their constitution and origin; but such correlations are at present too obscure for useful discussion.

In the preparation of filter paper for quantitative work it is important to eliminate the ash constituents as far as possible, and this is effected by treatment with hydrofluoric and other acids—'Swedish' filter paper of good quality contains from 0·03–0·05 p.ct. ash constituents, and constitutes the purest form of cellulose with which we can deal.

Cellulose and Water. Cellulose Hydrates.—All vegetable structures in the air-dry condition retain a certain proportion of water—or *hygroscopic moisture*, as it is termed—which is readily driven off at 100°, but reabsorbed on exposure to the atmosphere under ordinary conditions. The mean percentage of this 'water of condition' varies from 6 to 12 in the several celluloses; and in any particular cellulose will vary on either side of this mean number to the extent of 1–2 p.ct. with the extreme range of ordinary atmospheric conditions of temperature and tension of aqueous vapour.

The authors have made experiments on the 'drying' of celluloses in a current of carbonic acid gas. The 'hygroscopic moisture' (6–8 p.ct.) is rapidly driven off from the air-dry fibrous celluloses at 90–100°, and there is a further small loss of water (1 p.ct.) on raising the temperature to 180°. The loss at 100–120° is 0·5 p.ct.; after that the loss is slow and probably due to decomposition. Gelatinous celluloses in the form of films (see p. 28) when dried at 90–100° also show a further loss, but much greater at higher temperatures. Thus in one experiment an air-dry film lost 8·6 p.ct. on drying at 100°; an additional 3·9 p.ct. on raising the temperature to 160°.

In an earlier age of the science the question might have been discussed whether this absorption and retention of water is a chemical or physical phenomenon; but this is rather a question of terms. The main points to be noted are (1) the property of attracting water is a property of the cellulose substance itself, and is not in any way dependent upon the form in which it occurs. The amorphous modifications of the celluloses obtained by solution and reprecipitation in various ways (*infra*) are equally 'hygroscopic.'

(2) The phenomenon is definitely related to the presence of

OH groups in the cellulose molecule, for in proportion as these are suppressed by combination (with negative radicles to form the cellulose esters) the products exhibit decreasing attractions for atmospheric moisture. It is to be noted that some of these synthetical derivatives are formed with only slight modifications of the external or visible structure of the cellulose, of which, therefore, the phenomenon in question is again shown to be independent.

(3) The 'condition' of the fibre-substance in respect of hygroscopic moisture is an important factor of such properties of fibre as make up its spinning qualities; it also seriously affects the tensile strength of papers and cellulose textiles.

(4) A study of the hydration and dehydration phenomena of the celluloses indicates an unbroken continuity in the series of cellulose-water compounds—or cellulose hydrates; of which series the 'water of condition' or hygroscopic moisture of a cellulose represents the final terms.

The proportion of water held by the celluloses in an atmosphere saturated with aqueous vapour is necessarily very much greater than in the ordinary atmosphere, partially saturated at the same temperature. (See H. Müller, Pflanzenfaser, p. 3.)

The 'moisture of condition' is a factor of some moment, first in the buying and selling of fibrous products, and secondly in the processes by which they are worked up (spinning, and 'finishing').

(1) In a delivery of 100l. value of a fibrous material, e.g. paper pulp or half stuff, the ordinary variations in the atmospheric moisture may occasion a difference of 1l. to 2l. in the value. It is important, therefore, to have a normal standard of reference. In the case of wood pulp or cellulose in which there is a large commerce it is customary to fix this at 10 p.ct., which means that 100 of air-dry pulp give 90 'dry' at 100° C.

If, therefore, in any test the percentage of dry pulp is estimated at any figure, the corresponding percentage of 'normal' *air-dry* pulp (10 p.ct. Aq) is obtained by adding $\frac{1}{9}$ to the percentage of dry pulp.

(2) *Cotton-spinning* is carried on under special and carefully regulated conditions of temperature and atmospheric moisture,

which have been arrived at as the result of accumulated observation and experience. Raw cotton, however, is not by any means a pure cellulose, and the spinning properties of the fibre are to a certain extent conferred by the substances associated, in admixture or combination, with the cellulose. There is no doubt, however, that the physical properties of the cellulose are largely modified by its water of condition; and the fine adjustments of these ultimate fibres to the conditions of the spinning frame, more especially in regard to the drawing and twisting, largely depend upon the maintenance of an 'optimum' of hydration of the cellulose.

(3) *Finishing processes—textiles and paper.*—The 'finish' of textiles and papers for the market is of very various kinds. The last operations are those of closing and 'surfacing,' and consist of the mechanical treatments of beetling, mangling (textiles), calendering, and glazing (textiles and paper).

The finish is considerably affected by the condition of hydration of the fibre, and this is affected by the method of drying up (air-drying or hot-drying) and the amount of moisture present in the fibre when submitted to the mechanical treatments. The operation of causes of this kind is necessarily somewhat obscure. The student should address himself to the work of observation of the phenomena of hydration of the celluloses, studying all the conditions which affect, and the changes which accompany, the loss and gain of water.

It is evident from a very superficial examination of the plant world that the celluloses originate in the gelatinous form, i.e. in a condition of extreme hydration. Hydrates of identical characteristics are obtained on precipitating cellulose from solutions in the several special solvents to be subsequently described. These hydrates differ in certain respects from the anhydrous or dehydrated celluloses; thus they dissolve in strong nitric acid, and in solutions of the alkaline hydrates of moderate concentration, they also are more readily attacked (hydrolysed) by boiling dilute acids and alkalis. It is necessary to keep this in view in regard to the determination of cellulose in fresh tissues. A previous dehydration of the tissue by air-drying or by long immersion in alcohol confers upon the cellulose a much greater resistance to hydrolytic actions, with the effect of increasing the proportion of cellulose surviving the treatments previously described as necessary for the elimination of the non-cellulose constituents.

Some more important aspects of these phenomena are dealt with in a paper upon

The Hydration of Cellulose (J. Soc. Chem. Ind. 4).—In an investigation of the celluloses of green fodder plants the authors showed that by a preliminary artificial dehydration—by long immersion in alcohol—the quantity of cellulose isolated by the usual process of alkaline hydrolysis and oxidation was considerably increased.

The following numbers obtained with a crop of oats may be cited as typical.

Percentage Cellulose Isolated.

	(a) Directly	(b) After alcoholic dehydration	Difference
Leaves	28·2	35·4	7·2
Stems	29·5	34·5	5·0

It is a matter of ordinary observation that the maturing of vegetable tissues is attended by loss of water, and it is clear from these results that the growing plant contains hydrated modifications of cellulose, which by mere dehydration are converted into the more resistant forms. It must also be recognised that the line of cellulose has to be drawn in an arbitrary manner. Products which are the residues of treatments of a certain degree of intensity must be so defined, and are not to be regarded as chemical individuals in the strict sense of the term.

The hydrates of cellulose generally react with iodine in aqueous solution, giving an indigo-blue colouration. They also exhibit an increased 'affinity' for those colouring matters which dye cellulose directly.

In all the more essential properties, however, no distinction can be drawn between the celluloses and their hydrated modification.

Solutions of Cellulose.—Cellulose is insoluble in all simple solvents, water included. In presence of certain metallic compounds, however, it combines rapidly with water, forming the gelatinous hydrates just described, which finally disappear in solution in the water. Of these solvents of cellulose the simplest is (1) ZINC CHLORIDE IN CONCENTRATED AQUEOUS SOLUTION (40 p.ct. $ZnCl_2$). The solution process requires the aid of heat (60-100°), and may be carried out as follows : 4-6 pts. $ZnCl_2$ are dissolved in 6-10 pts. water and one pt. cellulose (cotton) stirred in till evenly moistened. The mixture is set aside to digest at a gentle heat. When the cellulose is gelatinised the solution is completed by exposure to water-bath heat, stirring from time to time and renewing the water which evaporates. In this way, a homogeneous syrup is obtained. This solution is employed in the arts for making cellulose threads or filaments which are carbonised for use in the incandescent electric lamp; the 'carbon' so obtained having a sufficient resistance to mechanical strain with the suitable degree of electric conductivity (resistance) for the requirements of the lamp. In preparing the cellulose thread the viscous solution is allowed to flow from a narrow orifice into alcohol which precipitates a hydrate—a hydrated cellulose-zinc-oxide—of sufficient tenacity for manipulation as a thread. It is freed from zinc oxide by digestion in dilute hydrochloric acid and copious washing. The cellulose zinc chloride solution is also precipitated by water, retaining a much larger proportion of water (of hydration). After thorough washing and drying a product is obtained retaining from 18-25 p.ct. ZnO ; the variation in the proportion of ZnO to cellulose, no doubt, corresponding with variations in molecular weight of the latter, and these depending upon the molecular condition of the original cellulose and the conditions of the solution process.

(2) ZINC CHLORIDE AND HYDROCHLORIC ACID.—If the $ZnCl_2$ be dissolved in twice its weight of aqueous hydrochloric acid (40 p.ct. HCl) a solution is obtained which dissolves cellulose rapidly in the cold. This alternative process has certain advantages over the preceding, and is useful in laboratory investigations. So far it has received no industrial applications. It is to be noted that the cellulose dissolved in this reagent undergoes a gradual lowering of molecular weight (hydrolysis).

This process of dissolving cellulose is of value in the investigation of fibrous products in the laboratory in cases where an acid solvent is preferable, and where it is necessary to avoid heat.

If to the solution of pure cotton cellulose in this reagent bromine be added in quantity sufficient to colour the solution, the colour persists for a lengthened period, showing that there is no absorption of the bromine, and that, therefore, there are no $C=C$ groups in the cellulose molecule. With the lignocelluloses (see p. 138), which are also soluble in this reagent, and are known by other reactions to contain $C=C$ groups, there is considerable absorption of bromine.

It is also noteworthy that if this solution of cellulose be coloured with CrO_3, it persists for some time in the unreduced state. There cannot, therefore, be any free $CO.H$ groups in the cellulose molecule, and the observation rather throws doubt on the existence of such groups in an 'acetal' form—$CH{<}^{OX}_{OX}$.

(3) AMMONIACAL CUPRIC OXIDE.—The solutions of the cuprammonium compounds generally, in presence of excess of ammonia, attack the celluloses rapidly in the cold, forming a series of gelatinous hydrates which finally pass into solution. The solutions of the pure cuprammonium hydroxide are more active in producing these effects than the solutions resulting from the decomposition of a copper salt with excess of ammonia. Two methods are in common use for the preparation of these solutions, which should contain :

| 10–15 p.ct. | . . . . | Ammonia (NH_3) |
| 2–2·5 ,, | . . . . | Copper (as CuO) |

(1) To a solution of a cupric salt, ammonium chloride is added, and then sodium hydrate solution in sufficient excess; the blue precipitate is thoroughly washed upon a cloth filter, squeezed, and re-dissolved in ammonia solution of 0·92 sp.gr.

(2) Thin sheet copper is crumpled up, placed in a glass cylinder and covered with strong ammonia. Atmospheric air is drawn by aspiration so as to bubble through the liquid column at such a rate as to amount per hour to about forty times the volume of liquid used. In about six hours a solution is obtained of the composition given above (C. R. A. Wright).

Under the latter conditions the action of the solution upon the cellulose may be made simultaneous with its production. For this the cellulose and metal are mixed together as intimately as possible and exposed as described to the action of aqueous ammonia and oxygen.

There are various ways of accelerating the preparation of the cuprammonium solution from the metal. Thus, as compressed oxygen is now an ordinary commodity it is easy to substitute the pure gas for the atmospheric mixture, with the result that the volume of gas passing through the solution may be considerably reduced, and therefore the loss of ammonia lessened.

The oxidation of the copper is facilitated by contact with a metal which is 'negative' to the copper in presence of ammonia; or this differential disposition of the copper to be attacked may be more directly attained by means of the electric current, the copper to be attacked being brought into conducting connection with the negative, and the second metal with the positive pole of a battery—the latter being inserted in a porous pot within the alkaline liquid (Hime and Noad, English Patent 7716/89).

The solutions of cellulose in cuprammonium are of little stability, the cellulose being readily precipitated by the addition of neutral dehydrating agents such as alcohol, sodium chloride (and other salts of the alkalis), and even sugar. From a study of these solutions, indeed, Erdmann concluded (J. Pr. Chem. 76, 385) that they were not solutions of cellulose in the strict sense of the term, the cellulose being rather gelatinised and diffused through the solution as a highly attenuated (hydrated) solid of this description. Cramer, on the other hand, showed by osmotic experiments that this inference was unfounded and that the solution of the cellulose may be regarded as complete. According to modern views on the subject of solution generally, and the solution of colloids in particular, the lines drawn by the older investigators of these phenomena are of arbitrary value; gelatinisation being expressed as a continuous series of hydrations between the extreme conditions of solid on the one side and aqueous solution on the other. This point will be further considered later on.

The evidence goes to show that the solution process, though not the result of an oxidation of the cellulose—such as would be attended by reduction of CuO—is attended by a disturbance of the 'balance of oxidation' of the cellulose molecule. By prolonged contact with the cuprammonium the cellulose does in fact appear to be oxidised (to oxycellulose) (Prudhomme, J. Soc. Dyers and Col., 1891, 148).

The ammonia also undergoes oxidation, and the cuprammonium solutions, after keeping, will be found to contain a considerable quantity of *nitrite* (ibid.). Cotton cellulose does not appear to be hydrolysed by the process of solution, that recovered from the solution by precipitation by acids &c. having approximately the same weight as that of the fibre originally dissolved.

There are celluloses, on the other hand, which are partially hydrolysed, and when reprecipitated the cellulose recovered is found to be in defect, and the solution to contain dissolved carbohydrates.

Further investigation of these points is much needed, i.e. quantitative determination of the oxidation and hydrolysis of the several celluloses under treatment with the cuprammonium reagent. The evaporation of the cuprammonium solutions of cellulose upon glass surfaces gives a film of the mixed cellulose-cupric hydrate, but of little tenacity. It will appear as we proceed that high tensile strength of a film obtained from solutions of cellulose compounds indicates a relatively high molecular weight, and conversely, a brittle product is evidence that in forming the compound the molecular weight or aggregation of the cellulose has been lowered.

According to recent investigations of E. Gilson (Chem. Centr. 1893, ii. 530) cellulose may be crystallised from its solution in cuprammonium. If such solution is left to stand in a loosely closed vessel the ammonia escapes, cellulose being precipitated together with hydrated copper oxide. On removing the latter by treatment with hydrochloric acid, the cellulose is stated to remain in the form of nodular crystals. It is also stated that when sections of cellulosic tissues are allowed to remain for some time in contact with the reagent, then gradually washed with ammonia and water, the interior of the cells are found to contain the cellulose in crystalline form. This requires confirmation.

These cuprammonium solutions are, of course, deprived of their copper by digestion upon zinc, the latter metal replacing the copper in solution and, under carefully regulated conditions, without precipitating the cellulose, so that a colourless solution of the latter in zinc-ammonium-hydroxide results. Some of these solutions have been observed to be lævogyrate. Cotton cellulose in 1 p.ct. cuprammonium solution was found by Levallois to show a rotation of $-20°$; the rotation, however, is not constant, but varies with the concentration and the ratio of cupric oxide to cellulose in the solution. These observations have been called in question by Béchamp, but reaffirmed by the former observer, and apparently on sufficient evidence.

On adding a solution of lead acetate to these solutions of cellulose a precipitate is obtained of a compound of cellulose with lead oxide, but of variable composition; the compound $n(C_6N_{10}O_5.PbO)$ appears to result from the treatment of the ammoniocupric solution with finely divided lead oxide.

This property of gelatinising and dissolving cellulose has been taken advantage of in important industrial applications of the cuprammonium compounds. Vegetable textile fabrics passed through a bath of the cuprammonium hydroxide are 'surfaced' by the film of gelatinised cellulose, which retains the copper oxide (hydrate) in such a way that it dries of a bright 'malachite' green colour. By this treatment the fibres are further compacted together, and the fabric acquires a water-resistant character. The presence of the copper oxide is also preservative against the attacks of mildew, insects, &c. If the fabrics are rolled or pressed together when in the gelatinised condition they become firmly welded together on drying, and a variety of compound textures are produced in this way.

These fabrics are sold under the style or description of 'Willesden' goods; the manufacture being in the hands of a company whose works are situated at Willesden. The company's processes are based on the patents of Drs. J. Scoffern and C. R. A. Wright (*q.v.*).

AMMONIACAL CUPROUS OXIDE.—According to M. Rosenfeld (Berl. Ber. 12, 954) a concentrated solution of cuprous chloride in ammonia dissolves cellulose rapidly.

The reaction of cuprammonium with cellulose, although identified with the name of Schweitzer, appears to have been first noticed by Mercer. He employed a solution of ammonia of 0·920 sp.gr. saturated at the ordinary temperature with the cupric oxide (hydrate) and diluted with three volumes of water. Mercer investigated the reaction in regard to the influence of the conditions of treatment, showing that it was retarded by the presence of salts, and hence that the solutions obtained by decomposing the copper salts with excess of ammonia were much less active than equivalent solutions of the pure hydrate. He also showed that the activity of the solution was considerably retarded by raising its temperature, becoming very slight at 100° F.

Mercer's favourite method of demonstrating the reaction consisted in applying a solution of cupric nitrate to cotton cloth in spots, decomposing the nitrate by plunging the cloth into a weak solution of caustic soda, washing to remove the alkali, partially drying—in the air at ordinary temperatures, and exposing the cloth to the vapour of ammonia. In this way the cellulose was fully acted upon in the portions containing the oxide. The demonstration is an interesting one, and should be repeated by the student.

Theory of Action of Cellulose Solvents.—The causes underlying the processes of dissolution of cellulose above described will become more apparent as we proceed in the discussion of its special chemistry. For the present it is sufficient to point out that they depend upon the presence in the cellulose molecule of OH groups of opposite function, basic and acid, and that the compounds formed with the solvents are of the nature of double salts.

Qualitative Reactions and Identification of Cellulose.—The properties of cellulose which we have already discussed afford the means of identifying it : that is (1) by reason of its resistance to the action of oxidising agents, to the halogens and to alkaline solutions it is obtained as a residue from the treatment of vegetable tissues by these reagents in succession ; (2) it is soluble to gelatinous or viscous solutions in the reagents above described—viz. $ZnCl_2.Aq$, $ZnCl_2.HCl.Aq$, and $Cu\begin{cases} NH_3-NH_4O \\ NH_3-NH_4O \end{cases}$, from which it is obtained by precipitation in the amorphous form or as a gelatinous hydrate. These hydrates react in many cases with iodine, giving a blue colouration ; the reaction is determined upon the original cellulose by simultaneous treatment with iodine and a dehydrating solution.

The most effective reagent is prepared as follows : zinc is dissolved to saturation in hydrochloric acid and the solution evaporated to 2·0 sp.gr. ; to 90 parts of this solution are added 6 parts potassium iodide dissolved in 10 parts water, and in this solution iodine is dissolved to saturation. By this reagent cellulose is coloured instantly a deep blue or violet.

A superficial examination usually suffices to identify cellulose in the mass, and an examination with the microscope establishes the histological characteristics of the substance. There are cases, however, in which distinctive tests require to be applied, and these will be selected in order of convenience. Thus, by means of the chemical tests, cellulose has been identified as a constituent of many animal tissues (see p. 87) ; in these cases, of course, it could be identified in no other way.

It will be seen as we proceed that a number of the properties of cellulose are common to many of the 'compound celluloses' which are widely distributed in the plant world ; these are, however, differentiated by the special reactions depending upon the compounds or groups with which the cellulose may be combined.

Lastly, the cellulose group proper includes a number of substances which are differentiated from the typical cotton cellulose in some specific property. These will be noted subsequently.

Compounds of Cellulose.—The chemical inertness of cellulose is a matter of every-day experience in the laboratory, where it fulfils the important function of a filtering medium in the greater number of separations of solids from liquids. The functions which it discharges in the plant world as well as the numberless uses which it subserves in the world of humanity are all referable to the predominance of these negative characteristics. Cellulose, however, is a poly-hydroxy- compound, and enters into a number of reactions characteristic of the alcohols. These reactions, and the products of synthesis resulting from them, we shall deal with in order, proceeding from the less to the more definite.

In a general way the inertness of cellulose may be compared with that of inorganic salts, more particularly those which result from the combination of the weaker acids and bases. Cellulose in reaction shows both acid and basic characteristics, and, as we shall see, these properties may be explained by proximity of its OH groups to CO and to CH_2 groups respectively within the molecule.

It appears, moreover, that these OH groups are in a condition of reciprocal suppression, requiring the application of powerful reagents or severe conditions to bring them into reaction.

This condition of its OH groups appears to be associated with the endothermic constitution or configuration of the cellulose molecule. There is a good deal of evidence physiological and chemical that the formation of cellulose is associated with the absorption of energy beyond what may be taken as normal to a saturated compound of the empirical formula $C_6H_{10}O_5$.

DILUTE ALKALIS AND ACIDS.—It has been shown that pure bleached cotton enters into reaction with the acids and basic oxides when plunged even into cold and highly dilute solutions of these compounds (Mills). In illustration of this point the following results of experiments may be cited :—

Reagent	Temperature	Time	Weight absorbed
H_2SO_4	4°	3 mins.	0·00495
HCl	,,	,,	0·00733
NaOH	,,	,,	0·02020

The molecular ratio of the absorption of the two latter—viz. 3HCl, 10NaOH—appears to hold good for a somewhat wide range of conditions; and it may be noted that the same ratio was observed for silk, though the observation can only be regarded as a coincidence.

These reactions of cellulose have been by no means exhaustively investigated; as our knowledge of the group of celluloses and of their differentiations one from the other is extended, it becomes necessary to institute a careful comparison in regard to this property of 'absorbing' reagents.

The Typical Cellulose and the Cellulose Group 17

An examination of the structureless cellulose regenerated from solutions of cotton as alkaline thiocarbonate (see p. 29) shows an important differentiation from the original cellulose (bleached fibre). This form of cellulose, after careful purification, was found to combine with the caustic alkalis in dilute solution, in much larger proportion; thus from solutions of 3·1 p.ct. Na_2O the cellulose removed, i.e. combined with, the alkali to the extent of 5·6 p.ct. of its weight. With the dilute acids, on the other hand, no increased combination was observed.

This phenomenon has been more recently studied from the independent standpoint of thermal equilibrium. It has been shown that when pure cotton is plunged into dilute solutions of the acids and alkalis, liberation of heat takes place (Vignon). The rise of temperature was found to be slow, and, under the conditions chosen for the experiments, ceases after the lapse of seven to eight minutes. The following are typical results in calories per 100 grms. of cotton.

	KOH	NaOH	HCl	H_2SO_4
Raw cotton	1·30	1·05	0·65	0·60
Bleached	2·27	2·20	0·65	0·58

It would appear from these results that cellulose has the properties of a feeble acid, and of a yet feebler base. From the comparative insignificance of the 'affinities' involved, it might be inferred that they could have but a small determining value in regard to the uses or applications of cellulose. Recent researches, however, have shown that the combinations of cellulose with colouring matters, i.e. the dyeing properties of the fibre-substance, are largely dependent upon a play of 'affinities' of this order and narrow range. Vignon concludes, in fact, from a careful and exhaustive survey of dyeing phenomena, including the action of mordants, that they depend chiefly upon the interaction of groups of opposite chemical function, viz.

basic and 'acidic,' present in the colouring matter or mordant and the substance with which it combines.

This explanation certainly covers a wide range of such reactions, but we shall find that the molecular constitution of the fibre-substance is also an important factor. This point will be discussed subsequently.

Capillary Phenomena.—The absorption and transmission of solutions by cellulose is attended by a number of special effects. Schönbein appears to have been the first to observe that strips of unsized paper, of which one end was placed in an aqueous solution, e.g. of a metallic salt, will absorb and transmit the water more rapidly than the dissolved salt, which is therefore 'filtered out'; further, that to the various salts, cellulose manifests varying degrees of resistance to transmission in solution. These phenomena have been further studied by Lloyd (Chem. News, 51, 51) for metallic salts,[1] and by F. Goppelsroeder (Berl. Ber. 20, 604) for various colouring matters; the results of their observations constituting the beginnings of a method of 'capillary analysis or separation.' The subject is comparatively new and not yet systematised, but the method is undoubtedly capable of considerable extension.

Contrasted with the relatively feeble attractions of cotton cellulose for the acids and bases of low molecular weight there are a number of cases of special combinations which take place in much higher proportions.

Thus the fibre removes a considerable quantity of barium hydrate on digestion with the solution; and from solutions of the basic salts of lead, zinc, copper, tin, aluminium, iron, chromium, &c. the fibre takes up considerable but variable proportions of the respective basic oxides. The formation of these compounds underlies the well-known processes of 'mordanting' practised by the dyer and printer of textiles. The theory of these processes will

[1] More recently by E. Fischer and Schmidmer, Lieb. Ann. 272, 156.

be found fully treated of in the text-books of these arts. We can only call attention to those properties which are common to the group of basic oxides capable of acting as mordants, viz. (1) they are all oxides of di- or polyvalent elements; (2) they form colloid or gelatinous hydrates; (3) their salts dissociate in solution into acid and basic salt; (4) they are soluble in the alkaline hydrates either directly or in presence of 'organic' hydroxy- compounds. Certain of the acid oxides of the metals are also removed by cellulose from solutions of these salts, but in relatively small proportion; of these the stannic compounds are most important from the point of view of application as mordants.

Amongst 'organic' acid bodies, tannic acid is conspicuous for its 'affinity' for cellulose. From aqueous solutions of tannic acid cotton fibre takes up as much as 7-8 p.ct. of its weight; and the process of mordanting with this compound is one of the most generally useful.

The combinations of cellulose with colouring matter open up a number of interesting problems. A colouring matter may be said to dye a fibre or substance when it forms with it a 'lake' compound, a lake being merely a pigment form of the colouring matter in which its essential physical properties are preserved. By recent investigation the properties determining lake formation have been shown to be definitely correlated with the molecular constitution of the colouring matter, i.e. more particularly with the nature and disposition of its chromogenic groups (NH_2 : SO_3H, $COOH$, NO_2, $N.OH$, and OH groups).

An excellent treatment of this subject will be found in two papers by C. O. Weber in the J. Soc. Chem. Ind. 10,896, 12,650, to which the student is referred. A discussion of these problems is outside the scope of this work. It may, however, be pointed out that, as of course the phenomena of dyeing depend upon reciprocal attraction, we may confidently expect that further investigation will lead to a correlation of the specific or selective attractions of cellulose for colouring matters, with its molecular constitution. It should be remembered that there are three factors of the problem: (1) the constitution of the colouring matter; (2) that of the substance with which it combines; and (3) the condition of the colouring matter in solution, and the causes which determine its transference to the solid with which it is brought into contact. Of these we have precise knowledge of the first only; the 'theory of solution' is a recent development, and the third factor is at present to be dealt with only speculatively; and of the constitution of the celluloses we have at present only a general knowledge.

The further investigation of these problems is therefore probably the most promising direction from which to approach the position of the actual molecular constitution of the celluloses.

The actions of dilute alkaline and acid solutions at higher temperatures are, of course, more pronounced. The mineral acids of concentration, equal to semi-normal at the boiling temperature, rapidly destroy in the sense of disintegrating cellulose fibres, producing an important molecular change in the cellulose itself. The modified cellulose is brittle and pulverulent, and will be more fully described as the product of the action of concentrated hydrochloric acid, viz. as hydro- or hydracellulose. The time required for completing this change varies of course with the temperature and the concentration of the acid. The acid treatments of cellulose textiles, which are necessary incidents of bleaching and dyeing operations, are carried out as a result of practical experience well within the limits of safety; such treatments being for the most part in the cold ($<70°$ F.) and at strengths of 0·5–2·0 p.ct. (HCl, H_2SO_4). In dyeing operations requiring an acid bath and the boiling temperature, 'free' mineral acids are as much as possible avoided, acetic acid being substituted. This acid is without sensible action on cotton.

The action of the acids in disintegrating cellulose structures is undoubtedly *hydrolytic*, and of the same order, for instance, as their action upon cane sugar. The 'inversion' of saccharose by boiling with the dilute acids is not, it must be remembered, a simple process of hydration; according to the usual equation,

$$C_{12}H_{22}O_{11} + H_2O = \underset{\text{Dextrose}}{C_6H_{12}O_6} + \underset{\text{Levulose}}{C_6H_{12}O_6},$$

these products of the hydrolysis being susceptible of 'condensations,' in which the reverse action is determined. On the other hand, it has been recently shown (Wohl) that the conditions may be so chosen as to exclude the latter, viz. by operating in the cold and in presence of a minimum of water, in which case we get the surprising result that the hydrolysis of the sugar is determined by $\frac{1}{5000}$ p.ct. of its weight of HCl.

Applying these considerations to the case of the more complex cellulose molecule it is easy to see that it may undergo a series of hydration changes, with attendant resolutions, without any change of its empirical formula. The disintegrating action of the dilute acids appears to be of this kind.

The action of the aqueous acids upon cellulose has been investigated by various observers, amongst others by Crace Calvert, Girard (Compt. Rend. 81, 1105), C. Koechlin (Bull. Mulhouse, 1888). The latter observer gives the results of a study of the limiting conditions of action of aqueous sulphuric acid at various degrees of concentration. What may be called the critical concentration of the acid lies between the limits of 60-80° B. Thus with the mixture of 3 vols. of concentrated acid and 8 vols. water —i.e. an acid of 69° B.—at the ordinary temperature, its action upon cotton does not become evident until after three hours' exposure. With an aqueous acid containing 100 grms. H_2SO_4 per litre and at 80° C. the first appearances of change in the cotton are noticed at the expiration of five minutes; after thirty minutes' exposure there is sensible disintegration; and the completion of the action, i.e. conversion into a friable mass of hydrocellulose, requires an exposure of 60 minutes' duration.

To alkaline solutions at high temperatures, cotton cellulose is, on the other hand, very resistant. Solutions of caustic soda of 1-2 p.ct. Na_2O are without sensible action upon cotton at temperatures considerably over 100°. The principal operations in the process of bleaching cotton and linen textiles consist in drastic alkaline treatments of this kind, whereby the 'non-cellulose' constituents of the fibre are for the most part saponified and removed in solution in the alkaline lye. The oxidation processes which follow—viz. treatment with the hypochlorites, permanganates, &c. in dilute solution—although they may be regarded as the bleaching processes proper, really accomplish very little beyond removing residues or by-products of the alkaline treatment. It is also evident that resistance to alkaline treatment is a very important condition in the everyday uses of cellulose textiles.

H. Tauss has recently investigated the action of alkaline solutions upon various celluloses at high temperatures (J. Soc. Chem. Ind. 1889, 913; 1890, 883). Purified cotton cellulose, digested with solutions of sodium hydrate of 3 p.ct. Na_2O three times in succession, is attacked and converted into soluble products in the following proportions, increasing with the temperature at which digested:

1 atm. pressure	. . . .	12·1 p.ct.
5 ,, ,,	. . . .	15·4 ,,
10 ,, ,,	. . . .	20·3 ,,

Strong aqueous solution of ammonia is without sensible action on cellulose until a very high temperature is reached. At 200° combination ensues, and the entrance of the NH_2 residue into the cellulose molecule is evidenced by the increased attraction of the product for colouring matters, approximating to that of the animal fibres. (L. Vignon.)

We have mentioned that digestion with 3 p.ct. solutions of soda (Na_2O) at high temperatures produces a certain conversion of cellulose into soluble substances. Solutions of 8 p.ct. (Na_2O) strength have been found to give the following results (Tauss, *loc. cit.*):—

1 atm. pressure	. . .	22·0 p.ct. dissolved
5 ,, ,,	. . .	58·0 ,, ,,
10 ,, ,,	. . .	59·0 ,, ,,

In connection with these observations it is to be noted that a process of estimating the cellulose in compound celluloses (wood) has recently been proposed (Lange, Zeitschr. f. Physiol. Chem. 14), and adopted by other observers, based upon the action of strong solutions of sodium hydrate at high temperatures upon the lignified tissue. It is assumed that the non-cellulose constituents of the woods (see p. 172) are exclusively attacked by the treatment: which, however, is by no means the case, as the results of Tauss (*loc. cit.*) sufficiently show. Quantitative results obtained by this method have, therefore, only a limited value; and, as estimations of 'cellulose,' are subject to large and variable errors.

CONCENTRATED SOLUTIONS OF THE ALKALIS.—Cold solutions of the alkaline hydrates of a certain concentration exert a remarkable effect upon the celluloses. Solution of sodium hydrate, at strengths exceeding 10 p.ct. Na_2O, when brought into contact with the cotton fibre, at the ordinary temperature, instantly changes its structural features, i.e. from a flattened riband, with a large central canal, produces a thickened cylinder with the canal more or less obliterated. These effects in the mass, e.g. in cotton cloth, are seen in a considerable shrinkage of length and width, with corresponding thickening, the fabric becoming translucent at the same time. The results are due to a definite reaction between the cellulose and the alkaline hydrates, in the molecular ratio $C_{12}H_{20}O_{10} : 2NaOH$, accompanied by combination with water (hydration). The compound of the cellulose and alkali which is formed is decomposed on washing with water, the alkali being recovered unchanged, the cellulose reappearing in a modified form, viz. as the hydrate $C_{12}H_{20}O_{10}.H_2O$. By treatment with alcohol, on the other hand, one half of the alkali is removed in solution, the reacting groups remaining associated in the ratio—$C_{12}H_{20}O_{10} : NaOH$. The reaction is known as that of Mercerisation, after the name of Mercer, by whom it was discovered and exhaustively investigated. Although, however, it aroused a good deal of attention at the time of its discovery, it remained for thirty years as an isolated observation, i.e. practically undeveloped. Recently, however, the alkali cellulose has been made the starting-point of two series of synthetical derivatives of cellulose, which must be briefly described.

An interesting account of Mercer's researches on this subject is given in 'The Life and Labours of John Mercer' (E. A. Parnell, London, 1886), a work which may be particularly commended to the young student.

From the points established by Mercer in connection with this reaction the following may be further noted:—

At ordinary temperatures a lye of 1·225–1·275 sp.gr. effects 'mercerisation' in a few minutes; weaker liquors produce the result on longer exposure, the duration of exposure necessary being inversely as the concentration. Reduction of temperature produces, within certain limits, the same effect as increased concentration. The addition of zinc oxide (hydrate) to the alkaline lye also increases its activity. Caustic soda solution of 1·100 sp.gr., which has only a feeble 'mercerising' action, is rendered active by the addition of the oxide in the molecular proportion, $Zn(OH)_2 : 4NaOH$.

The condition of the cotton also affects the result. The ordinary bleaching process, with its treatment with boiling alkaline lye under pressure, brings the cellulose into a condition relatively unfavourable, the best results being obtained by a preparatory treatment consisting of (1) boiling with water only, (2) bleaching in a warm bath (60–70° C.) of hypochlorite (bleaching powder) prepared with addition of lime.

In regard to the physical changes of the fibre-substance resulting from the treatment, the effects in the mass, i.e. in yarn or cloth, are seen in shrinkage of linear dimensions, with a corresponding increase in thickness. The percentage of shrinkage observed is 20–25. The 'mercerised' fabric shows an increase of strength, i.e. resistance to breaking strain, of from 40–50 p.ct. Another important feature of the 'mercerised' fabrics is an increased dyeing capacity. These changes of form and in properties were investigated by W. Crum (Chem. Soc. Journ. 1863).

The changes in the minute structure of the cell he showed to be similar to those which accompany the process of ripening —i.e. from the flattened riband form of a collapsed tube to the cylindrical form resulting from the uniform thickening of the cavity of the cell wall. Owing to this thickening the cavity of the cell is almost obliterated. Another effect of the alkali is to produce a peculiar spiral twisting of the fibre, which further explains the shrinkage of cloth in the process of mercerising; the shrinkage being in part due to the felting together of the twisted fibres, after the manner of wool fibres in the process of 'fulling' cloth.

CELLULOSE THIOCARBONATES.—When 'mercerised' cotton, or more generally an alkali-cellulose (hydrate), is exposed to the action of carbon disulphide at the ordinary temperature, a simple synthesis takes place, which may be formulated by the typical equation:

$$X.ONa + CS_2 = CS.\genfrac{}{}{0pt}{}{OX}{SNa}.$$

The best conditions for the reaction appear to be when the reagents are brought together in the molecular proportions:

$$\begin{array}{ccc} C_6H_{10}O_5 & 2NaOH & CS_2 \\ 162 & 2\times 40 & 76 \end{array} \quad [30\text{-}40 H_2O];$$

the second ONa group being in direct union with the cellulose molecule, which reacts, therefore, as an alkali cellulose. The resulting compound may therefore be described as an alkali-cellulose-xanthate. It is perfectly soluble in water, to a solution of extraordinary viscosity. The course of the reaction by which it is produced is marked by the further swelling of the mercerised fibre and a gradual conversion into a gelatinous transparent mass, which dissolves to a homogeneous solution on treatment with water.

To carry out the reaction in practice, bleached cotton is treated with excess of a 15 p.ct. solution of NaOH, and squeezed till it retains about three times its weight of the solution It is then placed in a stoppered bottle with carbon disulphide, the quantity being about 40 p.ct. of the weight of the cotton. After standing about three hours at ordinary temperatures, water is added sufficient to cover the mass, and the further hydration of the compound allowed to proceed spontaneously some hours (e.g. over night). On stirring, a homogeneous liquid is obtained, which may be diluted to any required degree.

Thus prepared, the crude solution is of a yellow colour, due to by-products of the reaction (trithiocarbonates). The pure compound is obtained either by treatment of the solution with saturated brine or with alcohol. It forms a greenish-white flocculent mass or coagulum, which redissolves in water to a colourless or faintly yellow coloured solution. Solutions of the salts of the heavy metals added to this solution precipitate the corresponding xanthates. Iodine acts according to the typical equation:

$$\text{CS}^{\text{OX}}_{\text{SNa}} + ^{\text{XO}}_{\text{NaS}}\text{CS} + I_2 = 2\text{NaI} + \text{CS.}^{\text{OX}\cdot\text{XO}}_{\text{S}-\text{S}}\text{CS.}$$

The compound, which may be described as a cellulose dioxythiocarbonate, is precipitated in the flocculent form; it is redissolved by alkaline solution, in presence of reducing agents, to form the original compound.

The most characteristic property of the cellulose xanthates is (a) their *spontaneous decomposition* into cellulose (hydrate), alkali, and carbon disulphide—or products of interaction of the latter. When this decomposition proceeds in aqueous solution, at any degree of concentration exceeding 1 p.ct. cellulose, a jelly or coagulum is produced, of the volume of the containing vessel. These highly hydrated modifications of cellulose lose water very gradually, the shrinkage of the 'solid' taking place symmetrically. The following observations upon a 5 p.ct. solution (cellulose), kept at the ordinary atmospheric temperature, will convey a general idea of the phenomena attending the regeneration of cellulose from the alkali xanthate. The observations were made upon the solution kept in a stoppered cylinder; after coagulation the solution, expressed from the coagulum of cellulose by spontaneous shrinkage, was removed at intervals. Original volume of solution, 100 c.c.

Time in days		Vol. of cellulose hydrate			Diff. from 100 c.c. = vol. expressed
Coagulation . . 8th day					
First appearance of liquid . . } 11th ,,					
16th ,,	•	98 c.c.	•	•	2 c.c.
20th ,,	•	83·5	•	•	16·5
25th ,,	•	72·0	•	•	28·0
30th ,,	•	58·0	•	•	42·0
40th ,,	•	42·8	•	•	57·2
47th ,,	•	38·5	•	•	61·5

The shrinkage from a 5 p.ct. to a 10 p.ct. coagulum of cellulose hydrate is therefore extremely slow and fairly regular; from 10-12 p.ct. there is considerable retardation; and at 12-15 p.ct. the coagulum may be considered as a hydrate, stable in a moist atmosphere. It follows from these observations that if a 10-12 p.ct. solution be allowed to coagulate spontaneously, the resulting cellulose hydrate will undergo very small shrinkage if kept in a moist atmosphere. These observations indicate the uses which can be made of the solution in preparing cellulose casts and moulds.

As regards the problem of hydration and dehydration of the cellulose there are, of course, other methods of approximately determining the 'force' by which the water molecules are held. It is a problem of wide significance, by reason of the important part played by such hydrates in the economy of plant life. Further investigations of the problem, therefore, by the various known methods are being prosecuted.

(*b*) *Coagulation by heat.*—The solution may be evaporated at low temperatures to a dry solid, perfectly resoluble in water. If heated at 70-80°, however, the solution thickens; and at 80-90° the coagulation, i.e. decomposition, is rapidly completed. If the solution be dried down at this temperature in thin films, it adheres with great tenacity to the surface upon which it is dried. On treatment with water, however, the cellulose film may be detached, and when freed from the by-products of the reaction the cellulose is obtained as a homogeneous transparent colourless sheet or film, of great toughness, which, on drying, hardens somewhat, increasing in toughness and preserving a considerable degree of elasticity. From the properties of the solution and of the cellulose regenerated from it, it will be readily seen that both are capable of extensive applications.

QUANTITATIVE REGENERATION OF CELLULOSE FROM SOLUTION AS THIOCARBONATE.—Very careful experiments have been made to determine the proportion of cellulose recovered from solution as thiocarbonate. Weighed quantities of Swedish filter paper were dissolved by the process, and the solutions treated as follows : (a) Allowed to 'solidify' spontaneously at 15–18°. (b) Coagulated more rapidly at 55–65°. (c) Sulphurous acid was added in quantity sufficient to combine with one-third of the alkali present in the solution—the resulting solution being colourless : this was then set aside to coagulate spontaneously. The regenerated celluloses were exhaustively purified, by boiling in sodium sulphite solution, digesting in acid, digesting in water, &c., and, repeating the treatments until pure, they were finally dried for some days at 60° and finished at 100°.

The following results were obtained :

	Weight of original cellulose	Weight of regenerated cellulose
(a)	1·7335	1·7480
(b)	1·7415	1·7560
(c)	1·8030	1·8350

The results show a net difference of 1·1 p.ct. (increase), a quantity which, for practical purposes, may be neglected. As, however, the empirical composition of the regenerated cellulose indicates hydration to $4C_6H_{10}O_5.H_2O$ (*infra*), and a corresponding gain of 2·7 p.ct., it appears that there is a slight hydrolysis of even this very pure form of cellulose. From subsequent observations (p. 61) it will appear that the hydrolysis falls upon an oxycellulose, probably present in all bleached celluloses.

The *cellulose regenerated* from the thiocarbonate differs from the original cellulose, so far as has been ascertained, in the following respects :

(1) Its *hygroscopic moisture*, or water of condition, is some 3-4 p.ct. higher, viz. from 9-10·5 p.ct.

(2) *Empirical composition.*—The mean results of analysis show C=43·3, H=6·4 p.ct., which are expressed by the empirical formula, $4C_6H_{10}O_5.H_2O$.

(3) *General properties*, in the main, are identical with those of the original, but the OH groups of this cellulose are in a more reactive condition. Thus this form of cellulose is acetylated by merely heating the acetic anhydride at its boiling point, whereas normal cellulose requires a temperature of 180°. (*Vide* Cellulose Acetates.)

As regards reaction in aqueous solution we may notice that it has a superior dyeing capacity, and also combines with the soluble bases to a greater extent: e.g. if left some time in contact with a normal solution of sodium hydrate it absorbs from 4·5-5·5 p.ct. of its weight in combination.

Towards the special solvents previously described it behaves similarly to the normal or fibrous cellulose; the solutions obtained are, however, more viscous and less gelatinous.

THEORETICAL VIEW OF THE THIOCARBONATE REACTION OF CELLULOSE.—The occurrence of this reaction, under what may be regarded as the normal conditions, proves the presence in cellulose of OH groups of distinctly alcoholic function. The product is especially interesting, as the first instance of the synthesis of a soluble cellulose derivative—i.e. soluble in water—by a reaction characteristic of the alcohols generally. The actual dissolution of the cellulose under this reaction we cannot attempt to explain, so long as our views of the general phenomena of solution are still only hypotheses. There is this feature, however, common to all the processes hitherto described, for producing an aqueous solution of cellulose (i.e. a cellulose derivative), viz. that the solvent has a saline character. It appears, in fact, that cellulose yields only under the simultaneous strain of acid and basic groups, and therefore we may assume that the OH groups in cellulose are of similarly opposite function. In the case of the zinc chloride solvents there cannot be any other determining cause, and the soluble products may be regarded as analogous to the double salts. The retention of the zinc oxide by the cellulose, when precipitated by water, is an additional evidence of the presence of negative or acidic OH groups; and, conversely, the much more rapid action of the zinc chloride in presence of hydrochloric acid indicates the basicity of the molecule, i.e. of certain of its OH groups. On the other hand, in both the cuprammonium and thiocarbonate processes there may be a disturbance of the oxygen-equilibrium of the molecule; and, although there is no evidence that the cellulose regenerated from these solutions respectively is oxidised in the one case, or deoxidised in the other, it is quite possible that temporary migration of oxygen or hydrogen might be determined, and contribute to the hydration and ultimate solution of the cellulose. But, apart from hypotheses, we may lay stress on the fact that these processes have the common feature of attacking the cellulose in the two directions corresponding with those of electrolytic strain; and it is on many grounds probable that the connection will prove to be causal and not merely incidental

The Typical Cellulose and the Cellulose Group   31

The thiocarbonate reaction more especially throws light on that somewhat vague quantity, the 'reacting unit' of cellulose. We use this term in preference to that of molecular weight; for the latter quantity can be determined only for bodies which readily assume the simplest of states, and which can be ascertained by physical measurements to be in that state; whereas in the case of cellulose the ordinary criteria of molecular simplicity are quite inapplicable.

We have formulated the synthesis of the thiocarbonate as taking place by the interaction of $C_6H_{10}O_5 : 2NaOH : CS_2$; or in approximate percentage ratio :

Cellulose : Alkali : Carbon Disulphide = 100 : 50 : 50;

or, again, in terms of the constituents estimated in the analysis of the product :

Cellulose : Alkali (Na_2O) : Sulphur = 100 : 40 : 40.

If now the crude product be precipitated from aqueous solution by alcohol or brine, and again dissolved and reprecipitated, the ratio changes to 100 : 20 : 20 ; and, through a succession of similar treatments, the ratio of alkali and sulphur to cellulose continually diminishes the product, however, preserving its solubility. In fact, no definite break has been observed in the continuous passage from the compound as originally synthesised to the regenerated cellulose (hydrate). It is clear, therefore, that the reacting cellulose unit is a continually aggregating molecule; and if in the original synthesis it appears to react as $C_6H_{10}O_5$, so in a thiocarbonate containing, e.g. only 4 p.ct. Na_2O, the unit is $10C_6H_{10}O_5$. There being, moreover, no ascertainable break in the series, we have no data for assigning any limiting value to the reacting unit under these conditions. All we can say is, that the evidence we have points to its being of indefinite magnitude; and we can see no *a priori* reason why it should not be so.

In discussing this reaction we have left out of consideration the part played by the water. It may be noted that a 1 p.ct. solution of cellulose (as thiocarbonate) will 'set' to a firm jelly of hydrate, of the volume of the containing vessel; and that even at 0·25 p.ct. cellulose, gelatinisation of the liquid occurs in decomposition. We have also pointed out that a hydrate containing only 10 p.ct. cellulose is a substantial solid which gives up water with extreme slowness.

Cellulose, therefore, affords conspicuous illustrations of the property which the 'colloids' have, as a class, of 'fixing' water, and of the modes in which this property takes effect. In regard to the causes underlying this peculiar relationship to water, we know as yet but little. It is to be noted that the group of colloids comprises bodies of very various chemical function, acids, bases, salts and compounds of mixed function, as in the complex carbohydrates and proteids; the only possible feature common to so varied a group would be that of molecular arrangement, favouring the aggregation of the molecules, together with those of water, to groups of indefinite magnitude. On this subject, however, conjectures must, for the present, do duty for a theory which can only be shaped by further investigation.

Cellulose Benzoates.—The alkali celluloses also react with benzoyl chloride, according to Baumann's method, to form the corresponding benzoates.

(a) *Mercerised cellulose.*—This form of alkali cellulose, treated with benzoyl chloride in the cold and in presence of excess of alkali, gives a mixture of products, the numbers obtained indicating that reaction occurs in the ratios,

$$\underset{\text{Cellulose}}{C_6H_{10}O_5} : \underset{\text{Benzoic acid}}{C_6H_5.COOH} \text{ and } C_6H_{10}O_5 : 2C_6H_5.COOH.$$

Within the limits of concentration, producing the specific 'mercerising' action—the lower limit being at about 12·5 p.ct. NaOH—the degree of benzoylation is inversely as the concentration of the alkaline solution. The fibrous benzoate produced under these conditions shows necessarily a much increased volume; examined microscopically the features of minute structure of the fibre are seen to be much accentuated. The hygroscopic moisture of the product is 2-3 p.ct. of its weight, i.e. from $\frac{1}{3}$ to $\frac{1}{2}$ that of the original cellulose. This weakened attraction for atmospheric moisture invariably attends the substitution of the OH groups in the celluloses by acid residues.

(b) *Soluble alkali celluloses.*—The hydrates precipitated from solution in the zinc chloride and cuprammonium solutions dissolve in solutions of the alkaline hydrates; and the benzoates obtained from these solutions, by treatment with benzoyl chloride, are curdy precipitates, which may be purified by solution in glacial acetic acid, filtering, and reprecipitating by water. Obtained in this way, the benzoates approximate in composition to $C_6H_8O_3\!\!<^{O.C_7H_5O}_{O.C_7H_5O}$. They melt at a high temperature to a clear liquid, which solidifies to a transparent resinous mass. By friction the compound becomes highly electric, a property common to the esters of cellulose.

The compound dissolves in acetic anhydride; on heating for some time at the boiling-point of the liquid, a partial replacement of the benzoyl by acetyl groups occurs, and at the same time a further acetylation of the cellulose.

The compound gives, on analysis, numbers corresponding with the empirical formula,

$$C_6H_6O.O.C_7H_5O.(O.C_2H_3O)_3.$$

This group of benzoates and mixed esters requires further and exhaustive investigation, as a study of their composition and constitution cannot fail to throw much light upon that of the parent molecule.

We now leave the alkali celluloses, and the synthetical

products which we have seen to be obtainable from them, to deal with those esters which are formed by direct synthesis with acid radicals. It will become evident, as we proceed with the description of the derivatives of cellulose, that it only yields, with any facility, such as result from reaction of its OH groups with mixed radicals. It will appear from the composition of the resulting esters about to be described that the empirical unit $C_6H_{10}O_5$ contains at least three OH groups ; of the remaining O atoms one certainly is carbonyl oxygen, though, as it exists in the cellulose complex, it manifests no 'outside' activity. It is brought into play, however, in a number of the decompositions of cellulose, and those determined by the non-oxidising acids are chiefly to be noted in regard to this point. It will become evident also that the molecular changes by which this group is set free are, in effect, decompositions, in the sense of a breaking down of the complex. The function of the remaining O atom is obviously a problem of moment in regard to the question of the constitution of cellulose, and it will appear that the solution of the problem presents considerable difficulties.

Cellulose Acetates.—The acetylation of hydroxy-compounds generally affords the simplest evidence as to their OH groups, their number and disposition or arrangement, in the molecule of the compound.

In the cellulose group, however, the problem is complicated (1) by the difficulties of preparation and purification of the acetates. The solutions are highly colloidal, and the ordinary criteria of purity of the products are wanting.

(2) By the difficulties of analysis ; the direct 'saponification' numbers often varying considerably from those obtained by distillation of the volatile acid (after saponification), and both numbers often failing to agree with the results of ultimate analysis.

(3) Some of the methods of acetylation certainly involve a change of molecular weight, and we have no criterion of the relation, in this respect, of the acetate to the original cellulose.

It must be understood, therefore, that the cellulose acetates about to be described are of undetermined molecular weight, and afford only an empirical expression of the number of OH groups in the undetermined unit $n.(C_6H_{10}O_5)$, which itself may vary under acetylation.

These considerations affect the value of deductions to be drawn from the composition of the acetates, as to the number of reactive OH groups in the unit molecule of cellulose.

On *a priori* grounds we should expect a maximum of 4OH in the unit $C_6H_{10}O_5$. For a long time, however, the 'triacetate' was considered to represent the highest degree of acetylation. This acetate, however, was obtained at an elevated temperature (180°) at which a variety of molecular complications are possible.

Acetates have been obtained by the following methods:—

(a) *Interactions of cellulose and acetic anhydride.*—On boiling cotton with acetic anhydride and sodium acetate, no reaction occurs. Heated at 180° in a sealed tube, in the proportion by weight of 1 of cellulose to 6 of the anhydride, the cellulose is converted into the triacetate (Schutzenberger). With the reagents in the proportion of 1 : 2, a mixture of lower acetates is formed. The latter are insoluble in glacial acetic acid ; the triacetate, on the other hand, is freely soluble. The solution is highly viscous, and passes with extreme slowness through filter paper. Filtration, however, is greatly facilitated by addition of benzene to the solution. The acetate also dissolves when heated with nitrobenzene, the solution gelatinising on cooling, even when highly dilute.

The cellulose acetates are easily saponified by dilute solutions of the alkaline hydrates, more rapidly in presence of alcohol (50 p.ct. vol.).

(b) *Cellulose and acetic anhydride in presence of zinc chloride.*—In presence of a relatively minute proportion of zinc chloride, cellulose and acetic anhydride react at 110–120°. The product, according to Franchimont, is the triacetate described above. According to later investigations, however, the acetylation proceeds further under these conditions, the numbers obtained indicating the formation of a tetracetate, and sometimes still higher numbers have been obtained. The following numbers show the quantitative relationship of the higher acetylated derivatives of a compound $C_6H_{10}O_5$:

			C	H	Acetic acid	Yields on saponification Cellulose
Triacetate	$C_{12}H_{16}O_8$	288	50·0	5·5	62·1	56·2
Tetracetate	$C_{14}H_{18}O_9$	330	50·9	5·6	72·7	49·1
Pentacetate	$C_{16}H_{20}O_{10}$	372	51·6	5·3	80·6	43·2

In dealing with the evidence as to the composition of these products, however, we must further remember that the formula $C_6H_{10}O_5$ has only a statistical value, i.e. that the molecule of cellulose is a complex aggregate; and if the molecule is partially resolved during acetylation, this may occur by way of hydrolysis or addition of OH groups. It is more than probable that, in presence of $ZnCl_2$, the acetylation is complicated in this way. It is always to be observed that, on pouring the product of the reaction into water, a fluorescent solution is obtained; and, further, that the cellulose regenerated from acetates prepared by this method is oxidised by cupric oxide in alkaline solution (Fehling's solution). These reactions indicate the liberation of the CO groups of the original cellulose during acetylation, and the reaction is not of such simplicity that we can draw any certain conclusions from the products as to the problematical O atom or atoms in the unit $C_6H_{10}O_5$.

(c) *Cellulose and acetic anhydride in presence of iodine.*—

The Typical Cellulose and the Cellulose Group 37

The addition of iodine in relatively small proportion ($\frac{1}{20}$ p.ct.) determines the solution of cellulose in acetic anhydride at 120–130°. By this method an acetate is obtained free from coloured by-products, and the yields of the product are remarkably uniform. In a series of experiments in which the proportions of cellulose and anhydride were considerably varied, the following yields were obtained per 100 parts cellulose:

176, 175, 177, 174;

the yield calculated for the triacetate being 177. The uniformity in these numbers, however, is somewhat illusory; as in certain cases the product is entirely soluble in acetone, and gives on analysis the numbers calculated for a triacetate; in other cases the product is resolved into a soluble fraction giving high saponification numbers (75 p.ct. acetic acid), and an insoluble fraction giving low numbers (48 p.ct. acetic acid; calc. for diacetate, 49). The evidence from these several processes is somewhat conflicting, as to the composition of acetates higher than the triacetate, and their relationship to the parent molecule.

(d) *Cellulose regenerated from solution as thiocarbonate.*—This form of cellulose has been found to react directly with acetic anhydride, under what may be considered the normal conditions. At 110–120° the cellulose is gradually dissolved, to a solution of quite extraordinary viscosity, which is so marked that the limit of concentration is not higher than 10 p.ct. of the acetate (5 p.ct. cellulose), beyond which point, i.e. the reaction is practically arrested. It is necessary, therefore, to use a very much higher proportion of anhydride to cellulose (20 : 1) than in the reactions previously described. The acetate thus obtained appears, from all its properties, to be a true derivative of cellulose. Thus it may be prepared in films of great tenacity and remarkable lustre; and the cellulose regenerated by saponification retains the film form, shows no tendency to be further hydrolysed by the alkaline solutions used for saponifying, and is unaffected by boiling with alkaline cupric oxide. The analyses of this acetate show satisfactory concordance of numbers with those calculated for a tetracetate : $n[C_6H_6O.(O.C_2H_3O)_4]$.

The specific gravity of this acetate is 1·210. It is soluble in acetone, methyl alcohol, glacial acetic acid, and nitrobenzene. It dissolves in concentrated nitric acid (as do the cellulose acetates generally), and is precipitated on dilution, apparently without change.

This compound is of 'critical' value in elucidating the constitution of cellulose. If the above formula be established by further and exhaustive investigation, the cellulose 'unit' must be $C_6H_6O.(OH)_4$; this is consistent with a cyclic arrangement of the carbon nuclei, and probably a symmetrical disposition of the OH groups. This question will be referred to subsequently.

Cellulose and Nitric Acid.—Cellulose Nitrates or Nitro-Celluloses.—The action of nitric acid on starch was investigated to some extent by Braconnot in 1833, who found that a very rapidly burning body was produced, and which was called xyloïdine. Pelouze further investigated this substance in 1838, and also similar bodies from paper, linen, &c. which he held to be identical with the one from starch. Schönbein is generally credited with the discovery of gun cotton in 1846. It appears to have been almost simultaneously discovered by Böttger, and also by Otto.

Whenever cellulose, in any form, is brought into contact with concentrated nitric acid at a low temperature a nitro-product, or a nitrate, is formed. The extent of the nitration depends upon the concentration of the acid ; on the time of contact of the cellulose with it, and on the state of the physical division of the cellulose itself.

Knop, and also Kamarsch and Heeren, found that a mixture of sulphuric acid and nitric acid also formed nitrates of cellulose; and still later (1847), Millon and Gaudin employed a mixture of sulphuric acid and nitrates of soda or potash, which they found to have the same effect.

Although gun cottons, or pyroxylines, are generally spoken of as nitro-celluloses, they are perhaps more correctly described as cellulose nitrates, for unlike nitro-bodies of other series, they do not yield, or have not as yet done so, amido-bodies on reduction with nascent hydrogen. Eder gives the following as general properties of the cellulose nitrates : (1) when warmed with alkaline solutions, nitric acid is removed in varying quantities dependent on the strength of the alkaline solutions employed; (2) treatment with cold concentrated sulphuric acid expels almost the whole of the nitric acid ; (3) on boiling with ferrous sulphate and hydrochloric acid, the nitrogen is expelled as nitric oxide ; the reaction is used as a method of nitrogen estimation in the cellulose nitrates ; (4) the alkaline sulphydrates, ferrous acetate, and many other substances convert the nitrates into ordinary cellulose.

Several well-characterised nitrates have been formed, but it is a very difficult matter to prepare any one in a state of purity, and without admixture of a higher or lower nitrated body.

The following are known : —

Hexa-nitrate, $C_{12}H_{14}O_4(NO_3)_6$,[1] gun cotton. In the formation of this body, nitric acid of 1·5 sp.gr. and sulphuric acid of 1·84 sp.gr. are mixed, in varying proportions, about 3 of nitric to 1 of sulphuric; sometimes this proportion is reversed,

[1] To represent the series of cellulose nitrates so as to avoid fractional proportions the ordinary empirical formula is doubled and the nomenclature has reference to this double molecule.

and cotton immersed in this at a temperature not exceeding 10° C. for 24 hours : 100 parts of cellulose yield about 175 of cellulose nitrate. The hexa-nitrate so prepared is insoluble in alcohol, ether, or mixtures of both, in glacial acetic acid or methyl alcohol. Acetone dissolves it very slowly. This is the most explosive gun cotton. It ignites at 160-170° C. According to Eder the mixtures of nitre and sulphuric acid do not give this nitrate. Ordinary gun cotton may contain as much as 12 p.ct of nitrates soluble in ether-alcohol. The hexa-nitrate seems to be the only one quite insoluble in ether-alcohol.

Penta-nitrate, $C_{12}H_{15}O_5(NO_3)_5$. This composition has been very commonly ascribed to gun cotton. It is difficult, if not impossible, to prepare it in a state of purity by the direct action of the acid on cellulose. The best method is the one devised by Eder, making use of the property discovered by De Vrij, that gun cotton (hexa-nitrate) dissolves in nitric acid at about 80° or 90° C., and is precipitated, as the penta-nitrate, by concentrated sulphuric acid after cooling to 0° C.; after mixing with a larger volume of water, and washing the precipitate with water and then with alcohol, it is then dissolved in ether-alcohol, and again precipitated with water, when it is obtained pure.

This nitrate is insoluble in alcohol, but dissolves readily in ether-alcohol, and slightly in acetic acid. Strong potash solution converts this nitrate into the di-nitrate $C_{12}H_{18}O_8(NO_3)_2$.

The tetra- and tri-nitrates (collodion pyroxyline) are generally formed together when cellulose is treated with a more dilute nitric acid, and at a higher temperature, and for a much shorter time (13-20 minutes), than in the formation of the hexa-nitrate. It is not possible to separate them, as they are soluble to the same extent in ether-alcohol, acetic ether, acetic acid, or wood spirit.

On treatment with concentrated nitric and sulphuric acids, both the tri- and tetra-nitrates are converted into penta-nitrate and hexa-nitrate. Potash and ammonia convert them into di-nitrate.

Cellulose di-nitrate, $C_{12}H_{18}O_8(NO_3)_2$, is formed by the action of alkalis on the other nitrates, and also by the action of hot dilute nitric acid on cellulose. The di-nitrate is very soluble in alcohol-ether, acetic ether, and in absolute alcohol. Further action of alkalis on the di-nitrate results in a complete decomposition of the molecule, some organic acids and tarry matters being formed. (See *infra*.)

The above account of the cellulose nitrates may be regarded as representing a fair digest of the extensive literature of the subject, so far as regards the composition and properties of the more important and definite products. A better grasp of the relationship of these products to one another and to the parent molecule will be obtained from the researches of Vieille (Compt. Rend. 95, 132), a short account of which follows. From the title of this author's communication, ' Sur les degrés de la nitrification limites de la cellulose,' it may be concluded that it is a study of the nitrations of cellulose (cotton) under the condition of progressive variations, with the view of determining the maximum fixation of the nitric group cor-responding to such variations. The most important factor of the process is the concentration of the nitric acid, which was the variant investigated. The temperature was kept con-stant—11° C.—and the nitrating acid (nitric acid only) was employed in very large excess (100–150 times the weight of cellulose), so as to avoid disturbance of the results by rise of temperature or by dilution of the acid. The products were analysed by Schlœsing's method, and the analyses are expressed in cc. NO (gas) (at 0° and 760 mm.) per 1 grm. of substance.

Sp. gr. of acid	Composition (approximate)	Analysis of product cc. NO per 1 grm.	Properties of products
1·502 1·497	$NO_3H.\frac{1}{4}H_2O$	202·1 197·9	Structural features of cotton preserved; soluble in acetic ether; *not* in ether-alcohol $C_{24}H_{26}(NO_3H)_{10}O_{10}$
1·496 1·492 1·490	$NO_3H.\frac{3}{8}H_2O$	194·4 187·3 183·7	Appearances unchanged; soluble in ether-alcohol; collodion cotton $C_{24}H_{27}(NO_3H)_9O_{11}$ $C_{24}H_{28}(NO_3H)_8O_{12}$
1·488 1·483	$NO_3H.\frac{1}{2}H_2O$	165·7 164·6	Fibre still unresolved; soluble as above, but solutions more gelatinous and thready $C_{24}H_{29}(NO_3H)_7O_{13}$
1·476 1·472 1·469	$NO_3H.\frac{3}{4}H_2O$	140·5 140·0 139·7	Dissolve cotton to viscous solution; products precipitated by water; gelatinised by acetic ether; *not* ether-alcohol $C_{24}H_{30}(NO_3H)_6O_{14}$
1·463 1·460 1·455 1·450	$NO_3H.H_2O$	128·6 122·7 115·9 108·9	Friable pulp; blued strongly by iodine in KI solution; insoluble in alcoholic solvents $C_{24}H_{31}(NO_3H)_5O_{15}$ $C_{24}H_{32}(NO_3H)_4O_{16}$

In regard to the time factor, or duration of exposure to the acid required to give the maximum number, this was in all cases controlled by observation. Thus with the acid $HNO_3.\frac{1}{2}H_2O$ (1·488 sp.gr.), after 48 hours the product was still blued by iodine, and gave 161 c.c. NO; whereas after 62 hours' exposure the iodine reaction was not obtainable, and the maximum number (165·7 c.c. NO) was obtained. At the slightly lower gravity 1·483, an exposure of 120 hours was necessary. At the still lower gravity when the cotton (nitrate) passes into solution, the maximum is very rapidly attained (5 minutes).

The highest nitrate obtained as above, with nitric acid only, is somewhat lower than when sulphuric acid is present. Under these latter conditions the author regards the highest nitrate obtainable as $C_{24}H_{15}(NO_3H)_{11}O_9$.

THERMAL CONSTANTS.—By calorimetric observation on the

process, it has been ascertained that the heat liberated is 11-12 cal. per unit of HNO_3 reacting. This 'heat of formation' is approximately equal to that which is observed in converting starch into the corresponding nitrates.

HEAT OF COMBUSTION.—The total combustion of *gun cotton* by free oxygen evolves heat equal to 2,300 cal. (H_2O of combustion *liquid*), or 2,177 cal. with the H_2O as gas or vapour, per 1 grm. of the compound. Collodion cotton gives the corresponding numbers 2,627-2,474 cal.; gun cotton exploded in confined space gives 1,071 cal. (H_2O of combustion liquid).

PRODUCTS OF COMBUSTION of gun cotton exploded in a closed vessel vary in relative amount and in composition with the 'density of the charge,' or the pressure developed at the moment of explosion. Thus the CO_2 and H increase with the density of charge; the CH_4 also, but, being present in very small ratio (0·0-1·6 p.ct. maximum), it may be neglected. The following equations may be taken as fairly representing the combustion of $2C_{24}H_{18}O_9(NO_3H)_{11}$, under varying conditions:

Density of charge
0·010 . . .	$33CO + 15CO_2 + 8H_2 + 21H_2O + 11N_2$
0·023 . . .	$30CO + 18CO_2 + 11H_2 + 18H_2O + 11N_2$
0·200 . . .	$27CO + 21CO_2 + 14H_2 + 15H_2O + 11N_2$
0·300 . . .	$26CO + 22CO_2 + 15H_2 + 14H_2O + 11N_2$

Under the ordinary conditions of explosion in firearms with maximum density of charge, the quantities of gas produced approximate more and more closely to the limit:

$$24CO + 24CO_2 + 17H_2 + 12H_2O + 11N_2.$$

Under 'explosion,' it will be seen that no nitric oxide or other nitrous gases are formed; but when a slower combustion takes place, with the products of combustion escaping freely under a pressure nearly equal to the atmospheric—as in a 'miss fire'—the percentage composition (by vol.) of the gases is

NO	24·7
CO	41·9
CO_2	18·4
H	7·9
N	5·8
CH_4	1·3
	100·0

(See Karolyi, Phil. Mag. 1863, 266; also Abel, Phil. Trans. 1866, 269; 1867, 181.)

INDUSTRIAL USES OF THE CELLULOSE NITRATES.—These products find a number of highly important uses both for destructive and constructive purposes. As far as these uses involve, or are based upon, essential properties of the products, they may be briefly noticed here.

EXPLOSIVES.—The products of which gun cotton—or other nitrated celluloses—is the essential constituent are of three main classes : (1) containing the nitrates only ; (2) the nitrates in admixture with inorganic salts containing oxygen 'available' for combustion, or aromatic nitro-derivatives, &c. ; (3) the nitrates in admixture with, or solution in, nitro-glycerin (blasting gelatine, ballistite, or cordite). An account of these modern explosives, with determinations of their constants of explosion, will be found in a paper by Macnab and Ristori, Proc. R. S. 1894, 56.

CELLULOID, XYLONITE, &c.—The lower nitrates are worked up with solvents of a special character (acetone, camphor, &c.), with or without admixture of various substances, into plastic masses, which are then cut and moulded into articles of most varied form and use.

COLLODION, COLLODION VARNISHES, COLLODION FILMS.—The lower nitrates, dissolved in ether-alcohol or other solvents (amyl acetate and benzene, &c.), form transparent solutions, which on evaporation leave the nitrate as a glass-clear film of considerable elasticity and tenacity. The products, both in solution and in the form of films, are applied in numerous directions, chiefly in connection with photography.

It is important to observe that these nitrates preserve in a remarkable degree the essential physical properties of the original cellulose, which will be most obvious by comparison of the above products with those obtainable with the cellulose regenerated from solution as thiocarbonate. But this is still better illustrated by the processes of converting the nitrates into a continuous thread, available as a textile material. This product is known as *artificial silk*. Various inventors have devised means for 'spinning' solutions of cellulose nitrates into thread, one of which may be briefly described as having reduced the operation to one of extreme mechanical simplicity. It is essential to the production of a thread of sufficient tensile strength—as directly obtained—to stand the strain of the drawing process, that the solutions employed contain a certain minimum proportion of the dissolved nitrate. Dr. Lehner, of Zurich,[1] after investigating the various problems involved, found that, whereas ordinary collodion containing such a proportion of the pyroxylin (10-12 p.ct.) in solution is unworkable under the prescribed conditions, the adding of dilute sulphuric acid causes a molecular change, and gives the solution the requisite fluidity. With such a solution the conversion into thread is effected as follows : The solution, carefully filtered and free from all bubbles, is caused to flow by way of glass tubes to a lower level, where it is delivered through a much narrowed opening with a steady constant flow. The shorter limb ending in this fine orifice is contained in a glass cell filled with water.

[1] See original German patent D.P. 58508/1890. The earlier processes of De Chardonnet (1885) and du Vivier (1889) must also be mentioned. See D.P. 38368 1885 and 46125/1888. Also Br. Pat. 2570 1889.

On emerging, therefore, the solution is at once coagulated to a transparent jelly, and of considerable toughness. On applying a slight pull to the jelly, grasped with the fingers or forceps, a thread is produced; and on fixing the end to a light wheel revolving at a definite rate, the thread is drawn off continuously of uniform diameter. Several threads being twisted together in the usual way of 'silk-throwing,' the artificial textile thread is produced. After being deprived of water of hydration the threads acquire the high white lustre of 'boiled-off' silk.

In this state, however, it is the explosive nitrate, containing 11–12 p.ct. N. To fit it for consumption, therefore, the 'silk' is 'denitrated' by treatment with ammonium sulphide in the cold. This process in no way affects the lustre of the thread, and when properly carried out gives a product not more inflammable than ordinary cotton.

The 'artificial silk' has been found to have a tensile strength equal to 70 p.ct. of that of the natural product, of the same degree of fineness. Its elasticity is inferior in about the same proportion; but it has a higher lustre and is produced at much less cost. It appears, therefore, capable of considerable industrial use.

OTHER DECOMPOSITIONS OF THE CELLULOSE NITRATES.—In addition to the explosive resolution into gaseous products, of these cellulose esters, they are susceptible of a more gradual process of decomposition, into which they pass spontaneously under certain conditions—yielding a complex of products, some of low molecular weight, e.g. carbonic, formic, oxalic, saccharic, and nitroxy-acids; others of closer relationship to the original cellulose, gummy acid bodies which have been described as belonging to the pectic series. Observations of these decompositions have been made by various chemists (Maurey, Béchamp, Kuhlmann, Pelouze, De Luca, Compt. Rend. **28**, 343; **37**, 134; **42**, 676; **59**, 363; **59**, 487;

Divers, Journ. Chem. Soc. [2], 1, 91), but have thrown but little light on the chemistry of cellulose.

A resolution of similar character is determined by a graduated treatment of the nitrates with the alkaline hydrates (solution). This has been investigated by W. Will, from the more theoretical point of view suggested by the title of the communication containing his results, viz. 'Ueber Oxybrenztraubensäure, ein neues Product des Abbaues der Cellulose,' Berl. Ber. 24, 400.

The process yielding this characteristic product, hydroxypyruvic acid, consisted in treating the ether-alcoholic solution of pyroxylin (with 11·2 p.ct. N) with a 10 p.ct. solution of sodium hydrate, shaking the two layers of solution together from time to time, until decomposition was complete; or setting aside for 24 to 30 hours, when it completes itself at ordinary temperatures. The alkaline solution is acidified and warmed, to complete the removal of the lower oxides of nitrogen, and treated with phenylhydrazine in presence of acetic acid (excess). The osazone of the ketonic acid, $COOH-CO-CH_2OH$, is thus obtained. The acid itself was also directly isolated from the original alkaline solution after neutralising, by precipitation as lead salt, and decomposing in the usual way with hydrogen sulphide.

The author's purpose in studying the reaction was the elucidation of the constitution of cellulose; and, although the results so far are too fragmentary for the drawing of definite conclusions, they indicate a direction in which the problem may be successfully attacked. It is obvious that progress in this direction must lie by way of processes of regulated dissection, and of these there are very few under sufficient control to be available. It is therefore to be hoped that this decomposition will be more fully investigated, especially as, from a private communication, we learn that the characteristic product is obtained in relatively large proportion, indicating a principal direction of cleavage of the cellulose molecule.

Attempts to arrive at the molecular weights of these compounds, benzoates and acetates as well as nitrates, by the method of Raoult have, so far, led to no result. The esters of cellulose appear to produce an abnormally large and, moreover, variable depression of the glacial point of acetic acid—which is a general solvent of these derivatives—such that no conclusion can be drawn from the observed depressions, as to the molecular magnitude of these compounds, in the undissolved condition; and if we interpret the depressions of freezing-point observed in the acetic acid solutions according to the usually accepted view, we must regard the molecules, when dissolved, as undergoing disaggregation or dissociation. There is no *a priori* objection to this view, and it appears, in fact, to be in harmony with many of the characteristics of cellulose in reaction, viz. those in which it resembles, to a certain extent, the inorganic salts.

Cellulose and Sulphuric Acid.—Cotton cellulose is rapidly attacked and dissolved by concentrated sulphuric acid. The initial product may, perhaps, be regarded as a cellulose sulphuric acid, but a rapid molecular disintegration ensues, and there results a series of sulphates of the general formula $C_{6n}H_{10n}O_{5n-x}(SO_4)_x$. The resolution of the cellulose molecule is a progressive phenomenon, and is accompanied by increase of dextro-rotation and reducing power (CuO) in the product.

The free acids are amorphous bodies, very hygroscopic, soluble in alcohol and water; on boiling the aqueous solution they are completely hydrolysed to glucose and sulphuric acid. The Ca, Ba, and Pb salts of the acids are obtained by neutralising with the respective oxides in aqueous solution and precipitating by alcohol. On boiling with water these salts lose one-half their sulphuric acid according to the equation

$$C_{6n}H_{10n-x}O_{5n-x}(SO_4H)_x + \tfrac{x}{2}H_2O$$
$$= C_{6n}H_{10n}O_{5n-\frac{x}{2}}(SO_4H)_{\frac{x}{2}} + \tfrac{x}{2}H_2SO_4.$$

(Hönig and Schubert, Monatsh. 6, 708; 7, 455.)

The reaction may be described, therefore, as a progressive hydrolysis of the cellulose through a series of dextrins, to a carbohydrate of minimum molecular weight. This transformation of cellulose to a sugar was established early in the century (Braconnot, 1819). Recent investigation has established the identity of this sugar with dextrose (rotation, $[\alpha] = 53\cdot 0$). The process of hydrolysis consists in the following stages: the cellulose (50 grms.) is dissolved in strong sulphuric acid (250 gr. H_2SO_4 + 84 gr. H_2O) in the cold, and the solution allowed to stand; diluted till the acidity equals 2 p.ct. H_2SO_4, and boiled 3 hours. The isolation of the dextrose in the crystalline form is accomplished in the usual way. (Flechsig, Zeitschr. f. Physiol. Chem. 7, 523.)

The reaction between cotton cellulose—as well as other celluloses of the cotton group (p. 79)—and sulphuric acid is, in regard to ultimate products, of the simplest character, resulting in their conversion into dextrose, and in quantitative proportion (Flechsig).

The reaction in the case of other celluloses, e.g. wood cellulose, is more complicated. The initial solution in the concentrated acid is dark coloured, and on diluting and boiling there is a considerable formation of insoluble products. Although dextrose is obtained as one of the ultimate products of the hydrolysis, it is only in relatively small quantity (Lindsey and Tollens, Annalen, 267, 371), and appears to be accompanied by other carbohydrates. It will be shown subsequently that the celluloses of this group are oxycelluloses, containing reactive CO groups and very readily condensing to furfural. The hydrolysis in such cases would, no doubt, be attended by condensations and other complications.

The subject has been recently and more exhaustively investigated by A. L. Stern (Thesis for D.Sc. Lond. Univ. 1894), and we give a short extract of the results which he obtained.

(1) *Composition of body produced by dissolving cellulose in sulphuric acid.*—In addition to the determination of the empirical ratios of the constituents in the products isolated as Ba salts, they were examined in solution for optical rotation and reduction

of cupric oxide. The former are expressed in terms of (d), and the latter in terms of dextrose reduction $K = 100$ (0·4535 grm. dextrose equivalent to 1 grm. CuO). (*a*) Solution of cellulose at 5°. (*b*) Solution at 15°. In both cases the compound obtained was $C_6H_8O_3(SO_4)_2Ba$; the compounds were without action on Fehling's solution; the 'opticity' varied directly with the temperature of the solution-reaction—viz. for (*a*) it was $+ 24°$; for (*b*) $+ 54°$. The yield of soluble Ba salt was 48 p.ct. of the theoretical; the residue of the cellulose remains associated with the $BaSO_4$, obtained on neutralising the acid liquid with $BaCO_3$.[1]

Hydrolysis of the product.—The compounds in solution were treated for 30 minutes at 100° in presence of free sulphuric acid (2 p.ct. on the solution). The acid products were isolated in each case as Ba salts. Compounds of identical formula were obtained—viz. $C_{18}H_{25}O_{12}(SO_4)_2Ba$. The yield amounted to (*a*) 95 p.ct., (*b*) 80 p.ct. of the total calculated. The remainder was converted into dextrose. The opticities of the products were different, viz. (*a*) for Ba salt $+ 25°$; (*b*) for Ba salt, $+ 75°$. The CuO reductions were, for (*a*), $K = 23·3$; for (*b*), $K = 18·1$.

(2) *Graduated hydrolysis of the disulphuric ester.*—The initial product of the empirical composition $C_6H_{10}O_3(SO_4)_2$ was then subjected to hydrolysis in graduated stages, the conditions being as before, and the stages being defined by the duration of the hydrolysis, the products being exhaustively investigated at periods of 7, 15, 20, and 30 minutes. The results are summarised as follows:

$C_6H_8O_3(SO_4H)_2$, when boiled with 2 p.ct. H_2SO_4, yields successively,

$$C_6H_8O_3(SO_4H)_2 \quad C_6H_9O_4.SO_4H$$
$$C_6H_8O_3(SO_4H)_2 \quad 3C_6H_9O_4.SO_4.H.$$

[1] This residue should be investigated.

No sugar (dextrose) is formed down to this stage, the result indicating a loss of sulphuric acid, and the formation of the monosulphuric ester, $C_6H_9O_4.SO_4H$, as the limit. This product, as the original disulphuric ester, is without action on Fehling's solution.

Subsequently the following were obtained as products of the further hydrolysis.

$$5C_6H_9O_4SO_4H \quad C_{12}H_{19}O_9SO_4H$$
$$2C_6H_9O_4SO_4H \quad C_{12}H_{19}O_9SO_4H.$$

This stage indicates further resolution of the monosulphuric ester; this takes place rapidly and is difficult to control.

The hydrolysis was then proceeded with for longer periods, 30–120 minutes. The results are thus summarised: sugar is formed, and acid products, with increasingly less barium and sulphuric acid. The sugar formed is dextrose. The following products were investigated:

$$6H_2SO_4.C_{12}H_{16}O_9 \quad 10H.SO_4C_6H_9O_4.2HSO_4C_{12}H_{19}O_9$$
$$H_2SO_4C_{12}H_{18}O_9 \quad HSO_4C_{12}H_{19}O_9$$
$$H_2SO_4C_{12}H_{18}O_9 \quad 4C_{12}H_{19}O_9SO_4H$$
$$H_2SO_4C_{12}H_{18}O_9 \quad 4C_{12}H_{19}O_9SO_4H.2C_6H_{12}O_6.$$

The corresponding Ba salts of these products contain more Ba than is necessary to saturate the acid group SO_4H. It is highly probable that one of the OH groups of the resolved cellulose molecule acquires acid functions.

The series of degradation products of the cellulose molecule are termed cellulose-sulphuric acids by the author; but this designation is misleading.

The most important features of this careful study of the molecular dissection of cellulose are, (1) the fact that the molecule can be very considerably resolved without freeing the aldehydic CO groups; (2) the differentiation of two of the OH groups of the C_6 unit, as having a superior basic or alcoholic function; and (3) that with the breaking down of the molecule, OH groups of the cellulose units are brought into play with acid functions.

It will be noted that up to this point we have been dealing with compounds of cellulose products obtained by synthetical reactions with acid and basic groups and with salts; in all of which the reacting molecule is maintained at or near its maximum weight (magnitude). We have mentioned incidentally, on the other hand, that the cellulose molecule, in the sense of the reacting unit, is a variable quantity; and that, while under certain conditions the tendencies are towards aggregation (thiocarbonate reaction), under others the tendency is towards a progressive disintegration. This is notably the case in the reaction with sulphuric acid just described, in which there is a perfectly graduated transition from the complex colloid molecule to the simple dextrose unit, a crystallisable solid of low molecular weight. These considerations lead up to the study of the

Decompositions of cellulose, which we shall find group themselves under the headings—

(*a*) Decompositions determined by the non-oxidising acids —the changes resulting from addition or subtraction of H_2O.

(*b*) Decomposition by oxidants, with attendant or secondary effects of hydrolysis and condensation.

(*c*) Decompositions by ferments; (*d*) by heat.

None of these decompositions of cellulose are of a simple character. Any aggregate change of composition can, of course, always be determined; as, however, we have no knowledge of the molecular magnitude and configuration, either of the parent molecule or of its derivatives—i.e. such as preserve the general characteristics of the celluloses—we are limited to the statistical study of these reactions, together with general inferences based upon their particular character.

(a) NON-OXIDISING ACIDS.—(1) *Sulphuric acid*, of 1·5-1·6 sp.gr. ($H_2SO_4.3H_2O$), produces the effects previously described, but in such a way as to be controlled within the earlier stages of molecular resolution. Unsized paper plunged into the cold acid, diluted to the above formula, is rapidly attacked, the paper becoming transparent owing to the swelling and gelatinisation of the fibres. The reaction quickly becomes one of solution; but if the paper be transferred, after short exposure, to water, the acid compound is at once decomposed and the resulting gelatinous hydrate of cellulose precipitated *in situ*. The product, after exhaustive washing and drying, is obtained as *parchment paper*. This modification of cellulose gives a tough translucent sheet, necessarily very much less absorbent than the original.[1]

The compound itself, from its resemblance to starch, has been termed *amyloid*. It is represented by the formula $n(C_{12}H_{22}O_{11})$, the semi-hydrate of $n(C_6H_{10}O_5)$. Like starch, the compound is coloured blue by iodine, and the joint action of iodine and sulphuric acid is frequently used in diagnosing cellulose. As a further result of the reaction, the product differs from cellulose, in containing active CO groups; it reacts with phenyl hydrazine salts, and is oxidised by CuO in alkaline solution.

Effects of a similar character are produced by treating cellulose with concentrated solutions of phosphoric acid and zinc chloride.

(2) *Nitric acid*, of 1·4 sp.gr., also produces (without oxidation) an effect of a similar character. A short immersion of unsized paper—e.g. filter paper—in the acid, followed by copious washing, has a considerable toughening action, due to superficial conversion of the fibres into a gelatinous hydrate.

---

[1] See Guignet on 'Soluble and Insoluble Colloidal Cellulose, and Composition of Parchment Paper,' Compt. Rend. 108, 1258.

These changes are marked by a shrinkage in linear dimensions of about $\frac{1}{10}$th: the tensile strength of the paper thus treated is about ten times that of the original. (J. Chem. Soc. 47, 183.)

(3) *Hydrochloric acid*, both in the form of gas and concentrated aqueous solution, rapidly disintegrates cellulose tissues. The product, obtained from cotton, after washing and drying, is a white powder which under the microscope is seen to consist of angular fragments of the original fibres. It has been termed *hydrocellulose* by Girard, who first described this product, and hydracellulose (Witz), the latter term being, perhaps, preferable. According to Girard, the product may be represented as $n(C_{12}H_{22}O_{11})$—as having, i.e. the same empirical composition as the above-described amyloid. From cellulose it also differs similarly to the latter, in the presence of free CO groups and the greater reactivity of its OH groups. Thus it reacts with acetic anhydride at its boiling point; the acetylation, however, does not proceed very far.

The product is *dissolved* by nitric acid (1·5 sp.gr.) without oxidation, and from the solution a series of nitrates are obtained: (1) the lowest nitrates, by spontaneous evaporation of the solution in their fibres; (2) higher nitrates, by precipitation with water; and (3) the highest nitrates, by precipitation with sulphuric acid.

From these properties it may be concluded that, in general molecular configuration, these derivatives are similar to cellulose, but are so modified that the typical reactions take place under much less extreme conditions.

The action of this acid we should expect to be one of dehydration; and, although the final product has the composition of a hydrate, there is every reason to regard the *hydration* as a secondary result, following the molecular rearrangement caused by the initial dehydration.

Although, therefore, the products resulting from the action of hydrochloric and sulphuric acids (1·55 sp.gr.) are identical in empirical composition, they are the very opposite in physical characteristics, and the actions of these acids certainly take very different courses.

It should be noted that the action of sulphuric acid at greater dilution (1·3 sp.gr.) approximates closely to that of hydrochloric acid, the product being a disintegrated and friable mass of the hydracellulose.

The non-oxidising acids generally produce similar results, the degree of action being proportionate to their hydrolytic activity.

A curious practical application of these processes of disintegrating cellulosic tissues may be noted in evidence of the fundamental chemical distinction of the vegetable (cellulose) fibres from those of animal origin (silk and wool). The latter are very resistant to the action of acids. From a wool-cotton fabric, therefore, the cotton is easily separated by soaking the fabric in dilute sulphuric acid, and, after removing the excess of acid, drying down on a hot floor. The disintegrated cellulose is then completely removed by dusting out, leaving the wool unaffected. A similar result is obtained with hydrochloric acid; or by treatment with certain chlorides which are dissociated, on heating, into hydrochloric acid and basic oxide—e.g. aluminium chloride or chlorhydrate.

On the other hand, the animal fibre-substances are extremely sensitive to the action of alkalis, to which, as we have seen, the celluloses are very resistant. The student should compare the constitution of the substances in question, so far as they have been elucidated, with that of cellulose, and for that purpose should read Berl. Ber. 1886, 850; J. Soc. Chem. Ind. 12, 426.

This activity being conspicuously feeble in the case of acetic acid, this acid has but little action upon cellulose, and therefore finds extensive use in the printing of cotton and linen fabrics.

Solutions of the mineral acids are extensively used in the 'souring' operations of the bleacher and dyer. They are usually employed cold, and the operation of souring is always followed by copious washing. Failure to remove the acid, even the last traces, results in disintegration or 'tendering' of the fabric on drying.

(*b*) **Decompositions of Cellulose by Oxidants.**—It has been already pointed out that cellulose is comparatively resistant to the action of oxidants; that most of the processes for isolating or purifying (bleaching) cellulose depend, *per contra*, upon the use of oxidising agents, which readily attack the 'impurities' with which it is combined or mixed in raw fibrous materials. The cellulose resists the action of these oxidising agents, and, further, withstands in a high degree the action of atmospheric oxygen. It is this general inertness of the compound which marks it out for the unique part which it plays in the vegetable world and in the arts.

It must be again noted that this high degree of resistance to hydrolysis (alkaline) and oxidation belongs only to cotton cellulose and to the group of which it is the type, and which includes the celluloses of flax, rhea, and hemp. A large number of celluloses, on the other hand, are distinguished by considerable reactivity, due to the presence of 'free' CO groups, and are therefore more or less easily hydrolysed and oxidised. The 'celluloses' of the cereal straws and esparto grass are of this type, and hence the relative inferiority of papers into the composition of which they enter. (J. Chem. Soc. 1894, 472.)

On the other hand, we have now to study those processes of oxidation to which it yields more or less readily.

A. OXIDATION IN ACID SOLUTIONS.—(1) *Nitric acid* (1·1–1·3 sp.gr.) attacks cellulose at 80–100°, at first slowly, then more rapidly, but tending to a limit at which the action again becomes very slow. This limit corresponds with the formation of a characteristic product of oxidation—*oxycellulose*. This substance, which is white and flocculent, when thrown upon a filter and washed with water, combines with the latter to form a gelatinous hydrate. It requires, therefore, to be rapidly washed with dilute alcohol. It amounts to about 30 p.ct. of the cellulose acted upon, the remainder being for the most part completely oxidised to carbonic and oxalic acids. On ultimate analysis it gives the following numbers:

$$\left.\begin{array}{ll} C & 43\cdot 4 \\ H & 5\cdot 3 \end{array}\right\} C_{18}H_{26}O_{16}.$$

It dissolves in a mixture of nitric and sulphuric acids, and on pouring into water, the nitrate $C_{18}H_{23}O_{13}(NO_3)_3$ separates as a white flocculent precipitate. From the low number of OH groups reacting with the nitric acid, it may be concluded that the compound is both a condensed as well as an oxidised derivative of cellulose. This oxycellulose dissolves in dilute solutions of the alkalis, and on heating the solutions they develop a strong yellow colour. Warmed with concentrated sulphuric acid it develops a pink colouration similar to that of mucic acid. The compound exhibits generally a close resemblance to the pectic group of colloid carbohydrates.

The by-products of this oxidation are carbonic and oxalic acids, together with the lower nitrogen oxides. The solution, examined at any stage, appears to contain traces only of intermediate products of oxidation of the cellulose. The reaction is divisible into the two stages: (1) the conversion of the cellulose into hydracellulose, evidenced by its breaking down to a fine flocculent powder; and (2) the oxidation of the hydracellulose.

The oxycelluloses resulting from this process differ from those formed by the action of CrO_3 (*infra*), in giving small yields only of furfural (2-3 p.ct.) on boiling with HClAq (1·06 sp.gr.). It is also to be noted that the carbon is higher than that of the oxycelluloses, giving large yields of furfural (p. 84). These points suggest that, side by side with oxidation, combination of the negative oxy-groups with the more basic groups of unattacked molecules takes place, giving derivatives of the nature of esters. And, indeed, the reaction may be even more complicated. It is clear, from the composition of the nitrate, that the proportion of basic OH groups is reduced to a minimum.

The reaction requires further systematic research in the light of our increased knowledge of the constitution of the simpler carbohydrates and the simple products of their oxidation.

(2) *Chromic acid*, in dilute solutions, attacks cellulose with extreme slowness; in presence of mineral acids oxidation proceeds more rapidly, but at ordinary temperatures is still very slow. The action is, therefore, easily controlled within any desired limit, the oxidation being in this case of course directly proportionate to the amount of CrO_3 presented to the fibre. The oxidation is accompanied by disintegration, and the insoluble product is an oxidised cellulose, or oxycellulose, the yield and composition of which bear a simple relation to the amount of oxidation to which the cellulose is subjected. Its properties are similar to those of the oxycellulose above described. It dissolves in a diluted mixture of sulphuric and hydrochloric acids (57 p.ct. H_2SO_4, 5·5 p.ct. HCl), and on diluting and distilling with HCl of 1·06 sp.gr., is decomposed with formation of furfural, $C_4H_3O.COH$, the yield of this aldehyde being proportionate to the state of oxidation of the product.

This is illustrated by the subjoined results of observations:

Weight of cellulose	CrO_3 employed	Yield of oxycellulose	Yield of furfural p.ct. of oxycellulose
4·7	1·5	93·0	4·1
4·7	3·0	87·0	6·3
4·7	4·5	82·3	8·2

(Berl. Ber. 26, 2520.)

The first effect of treatment with CrO_3 appears to be that of simple combination; reduction to the Cr_2O_3 then ensues, and the further deoxidation requires the presence of a hydrolysing acid.

From the statistics of the reaction it appears there is little 'destruction' of the cellulose; and, as the oxidation is not attended by evolution of gas (CO_2), we may assume that the reaction consists simply in oxidation with the fixation of water. A certain proportion of the products are dissolved by the acid solution, and of the insoluble residue (oxycellulose) a large proportion is easily attacked and dissolved by alkaline solutions. The product is no doubt, therefore, a mixture; and, indeed, it would be hardly conceivable that an aggregate like cellulose should be equally and simultaneously attacked.

The reaction is so perfectly under control that it must be regarded as giving a regulated dissection of the molecule of cellulose, and therefore is an especially attractive subject for exhaustive investigation.

The carbohydrates of low molecular weight are similarly oxidised by chromic acid, and the product of oxidation similarly resolved with formation of furfural.

It is to be noted with cellulose, as with the carbohydrates of low molecular weight, that by oxidation its equilibrium is disturbed in such a way that carbon condensation is easily determined. This fact is of physiological significance and will be referred to subsequently.

(3) *Of other acid oxidations* which have not been particularly investigated we may mention the action of Cl gas in presence of water, of hypochlorous acid, and of the lower oxides of nitrogen in presence of water. Generally the result of these treatments is similar: the formation of insoluble products having the properties of the oxycelluloses above described, and soluble products which are oxidised derivatives of carbohydrates of low molecular weight. These, however, are usually obtained in relatively small quantity.

Atmospheric oxidation of cellulose—if it could be proved to take place—would fall in this category, as cellulose surfaces under ordinary conditions of exposure would be found to be normally acid. From the evidence we have of the condition of paper and textiles of the flax group after centuries of exposure to ordinary atmospheric influences, we may conclude that the oxidation of the normal celluloses under these conditions is excessively slight.

B. Oxidations in Alkaline Solution.—(1) *Hypochlorites*, in dilute solution ($<$ 1 p.ct.) and at ordinary temperatures, have only a slight action upon cellulose; a fact of the highest technical importance, since hypochlorite of lime (bleaching powder) is the cheapest of all soluble oxidising compounds, and the most effective oxidant of the coloured impurities which are present in the raw cellulose fibres or formed as products of alkaline hydrolysis.

While the normal celluloses withstand these bleaching oxidations, there are many celluloses widely differentiated from the cotton type which are eminently oxidisable, and, at the same time, susceptible of hydrolysis. The 'celluloses' of esparto and straw are of this kind (see p. 84), and the economic bleaching of paper pulps prepared from these raw materials can hardly be expected to follow upon the same lines as that of 'rag' pulp (cotton and linen). A study of the factors involved in the process will be found in a paper entitled 'Some Considerations in the Chemistry of Hypochlorite Bleaching' (J. Soc. Chem. Ind. 1890). These factors are—in addition to temperature and concentration (Cl_2O) of the bleaching solution—the nature of the base in union with the hypochlorous acid, and its proportion to the acid. A knowledge of the operation of these factors will enable the bleacher to control a process which is usually carried out on an entirely empirical basis.

The resistance of cellulose to the action of these solutions necessarily has its limits, and when these are exceeded, the fibre-substance is oxidised and disintegrated, and an oxycellulose results. These effects are rapidly produced by the joint action of hypochlorite solutions and carbonic acid. The oxycellulose formed in this way acquiring the property of selective attraction for certain colouring matters—notably the basic coal tar dyes—its presence in bleached cloth is easily detected by a simple dyeing treatment consisting in immersing the oxidised fabric in a dilute solution (0·5-2·0 p.ct.) of one of these dye stuffs, e.g. methylene blue. Local over-oxidation may be diagnosed in this way with certainty, and bleachers' damages may be thus ascertained and often traced back to the operating cause in the light of this 'oxycellulose' test. (J. Soc. Chem. Ind. 1884.)

The oxycellulose or disintegrated fibre resulting from this process of oxidation differs but little in empirical composition from cellulose itself, probably owing to the fact that the more highly oxidised products are dissolved in the solution of the oxidant, which is, of course, basic. Its reactions indicate the presence of free CO groups, and it readily undergoes further oxidation by atmospheric oxygen, the oxidation being much accelerated by temperatures over 60°. The OH groups of this oxycellulose are also more reactive than those of the original cellulose, acetylated derivatives being obtained by boiling the product with acetic anhydride.

The facts in relation to the conversion of cotton cellulose into oxycellulose by the action of bleaching powder were first made known by George Witz in 1883 (Bull. Soc. Ind. Rouen, 10, 416; 11, 169).

Since when a number of papers have been published dealing with special aspects of the phenomena—theoretical and practical. Of these we may cite: Schmidt, Dingl. J. 250, 271; Franchimont, Rec. Trav. Chim. 1883, 241; Nölting and Rosenstiehl, Bull. Rouen, 1883, 170, 239; Nastjukow, Bull. Mulhouse, 1892, 493.

It is probable on many grounds that the oxidised products obtained from cellulose by the action of the hypochlorites in the manner described are mixtures of one or more oxycelluloses with residues of unoxidised cellulose. More recent investigation has led to the conclusion that the extreme product of oxidation is an oxycellulose of the empirical formula $C_6H_{10}O_6$, which is freely soluble in dilute alkaline solutions in the cold; and that cellulose oxidised by hypochlorite solutions is a variable mixture of this product, with hydracellulose, and unaltered cellulose. (Nastjukow.)

By drastic oxidation of cellulose by the oxyhalogen compounds—i.e. by treatment with chlorine or bromine in presence of alkaline hydrates—the molecule is entirely broken down to the simplest products. With bromine, i.e. hypobromite, some quantity of bromoform is obtained; carbon tetrabromide is also easily obtained and identified. (Collie, J. Chem. Soc. 65, 262.)

(2) *Permanganates.*—The permanganates in neutral solution attack cellulose but slowly, and they may therefore be usefully employed as bleaching agents. In presence of alkalis a more drastic oxidation is determined. The degree of oxidation is, of course, dependent upon the conditions of treatment. The following general account of a particular experiment and its results will illustrate its main features.

22·6 grms. cellulose, with 400 c.c. caustic soda solution; 50 grms. $KMnO_4$ added in successive small portions; temperature, 40–50°. Proportion of cellulose to oxidising oxygen, $2C_6H_{10}O_5 : 7O$.

The main products were:

(α) Oxycellulose	10·5 grms.,	approximately	50 p.ct,
(β) Oxidised carbohydrates in solution	3·5 ,,	,,	16 ,,
(γ) Oxalic acid	4·3 ,,	,,	20 ,,
(δ) Carbonic acid, water, and traces of volatile acids		·,	14 ,,

(α) The oxycellulose gelatinised on washing, and was similar to the product obtained by the action of nitric acid.

(β) The oxidised carbohydrate in solution resembled 'caramel' in appearance. The compound or mixture was precipitated by basic lead acetate, and isolated by decomposing the precipitate with hydrogen sulphide, filtering and evaporating. On distillation from hydrochloric acid, furfural was obtained in large proportion.

(3) *Extreme action of alkaline hydrates.*—When fused at

200-300° with two to three times its weight of sodium or potassium hydrates, cellulose is entirely resolved, the characteristic products being hydrogen gas, and acetic (20-30 p.ct.) and oxalic acids (30-50 p.ct.). Generally the reaction takes the same course as with the simpler carbohydrates, resolution of the cellulose into molecules of similar constitution no doubt preceding the final resolution, which appears to be an exothermic or explosive reaction.

Distinguished from the two groups of decomposition which we have now considered—viz. those determined (1) by hydrolytic agents, (2) by oxidising agents (under hydrolysing conditions)—are those of a more intrinsic character, determined rather by the addition or withdrawal of energy, than by reaction with outside molecules.

C. RESOLUTION BY FERMENT ACTIONS.—This group of decompositions of cellulose is necessarily a very wide one. In the 'natural' world of living organisms, of course, no structures are permanent; and although cellulose distinguishes itself by relative permanence and resistance to the disintegrating actions of water and oxygen, the differentiation in this respect is only a question of degree, and all cellulosic structures are subject to the law or necessity of redistribution.

The directions of redistribution are chiefly three . viz. (1) In the assimilating processes of the plant a cellulosic structure is broken down, reabsorbed into the supply of plastic nutrient material, and re-elaborated.

(2) Structures which have ceased to play a part in the general organisation of the plant are cast off and then exposed as 'dead' matter to the play of the redistributing agencies of the natural world. The processes of 'decay' take various forms according to the conditions to which they are exposed. The humus of soils, peat, lignite, and all forms of coal present various forms of the residual solid products of the decay of

cellulosic structures, the remainder having been dissipated and restored to the general fund of matter in circulation, in the gaseous form—viz. as CO_2 and CH_4.

(3) In the processes of animal nutrition plants and vegetable substances are, of course, most important factors. In the course of animal digestion the vegetable substances are attacked by the fluids of the alimentary tract and resolved into proximate constituents fulfilling the requirements of the organs of assimilation; and in addition to these decompositions, which are largely hydrolytic in character, more fundamental resolutions are observed in which the carbohydrate molecules are completely broken down, i.e. with formation of gaseous products.

Processes of the first of these three groups are well known to plant physiologists; tissues of a cellulosic character are specialised to serve as reserves of nutrient material, or reserve material is stored up within a cellulosic cell wall which requires to be broken down in order that the supply may be liberated. In the seeds of the Gramineæ, more especially the barleycorn, this process of reabsorption of a cellulosic tissue has been carefully studied, and there is no doubt that the process of breaking down the cellulose is due to the action of a special ferment—a cellulose-dissolving enzyme. It must be noted here that the celluloses susceptible of this simple form of hydrolysis are of very different constitution from the typical cotton cellulose, and the features of this differentiation will be discussed subsequently. Taking cellulose, however, as a general expression for the colloid carbohydrates, we may regard them as having the property of yielding to the hydrolytic activity of special enzymes.

As the student is now considering ferment actions he may take, in illustration of these general views, the alcoholic fermentation of dextrose. The main products of this decomposition—alcohol and carbonic acid—are so related that the decomposition may be explained as resulting from migration of oxygen in the one, and of hydrogen in the other and opposite direction within the molecule: a decomposition, therefore, of the electrolytic type. Nothing is known of the intermediate stages of the resolution of the $C_6H_{12}O_6$ into $2[C_2H_5OH + CO_2]$; the decomposition is rather of an explosive character, and we have so far no means of investigating its mechanism.

The sugars are, of course, not 'organised' as such into cellular tissue, but are built up into aggregates of specialised constitution. The reabsorption of such aggregates into the general circulation of nutrient material of the plant, as indicated under (1), is the result of *proximate* resolution of these aggregates. It must be borne in mind here that changes of this order have been brought to light by physiological and histological methods; and with very little regard to the chemistry of the changes or, indeed, the actual composition of the tissue substance. Later investigations are differentiating these tissue-substances altogether from the celluloses of the cotton type, and in reading this section the student may be reminded that in the classification of the celluloses (which follows later, p. 85), it will be shown that so-called 'celluloses' susceptible of hydrolytic degradation are of an inferior order of molecular aggregation—probably rather resembling that of the starch-dextrin series.

We may note here more particularly an important paper by Brown and Morris (J. Chem. Soc. 1890, 57, 458) upon the 'Germination of some of the Gramineæ.' It is generally known that the cell wall of endosperm cells containing nutrient substances, to be supplied to the embryo in its earliest stages of growth, are broken down, as a preliminary to the appropriation of the cell contents. The general mechanism of the process has been elucidated by the above observers, even to the localisation of the cellulose-dissolving enzyme (cyto-hydrolyst). This enzyme does not exist in the resting seed but is formed in the process of germination. From a cold water extract of an air-dried malt the enzyme is precipitated by alcohol. An extended investigation of its activity showed that it rapidly disintegrates the parenchymatous tissue of the potato, carrot, turnip, apple, beet, &c. The elegant methods of experiment pursued by the above authors are typical of chemico-biological work, and should be thoroughly mastered by the student.

In the second group of resolutions, constituting 'decay,' various micro-organisms play an important part. The extreme resolution takes place according to the equation

$$C_6H_{10}O_5 + H_2O = 3CO_2 + 3CH_4$$

(Hoppe-Seyler, Ztschr. Biol. 10, 401). This decomposition is determined by the amylobacterium, and may be taken as typical. Pure 'fermentations' of cellulose have, however, been but little investigated.

In the decay of plant structures we have to deal with a complex of compounds and with celluloses of very various character. Again, therefore, we can only treat of these processes in their broad and general features. These are, in the main, (1) complete resolution, of the kind described and formulated above; (2) a tendency in the precisely opposite direction, i.e. towards condensation of the carbon nuclei to still more complicated forms, accompanied by the splitting off of water. These processes are concurrent as they are in the decompositions by heat, about to be described. As visible and tangible results of this tendency to carbon accumulation we have the vast aggregations of peat, lignite, and coal in all its forms in the earth's crust, which are the residues of the flora of a past geological age. In the coal measures, moreover, there is abundance of gaseous carbon compounds also stored up. These being chiefly marsh gas and carbonic acid, the process of coal formation, in its earlier stages, appears to have been similar in all respects to those which we can observe around us as attending the decay of vegetable matter in the mass.

These decompositions are necessarily of a complex character, and are, no doubt, largely dependent upon the presence of nitrogenous substances. We may cite in illustration the disintegration of leaf parenchyma in the well known process of 'skeletonising.' Leaves of the poplar, pear, &c. are covered with water and set aside in a warm place (35-45°). In the course of a week or so the parenchymatous tissue is so far broken down and gelatinised that it is easily detached from the 'skeleton tissue' constituting the venation of the leaf. This is a cellulose, or rather lignocellulose (see p. 92), of the more resistant type; and the process affords a simple means of differentiating the cellulose group in regard to resistance to hydrolytic agencies.

The 'rot-steep' or retting of flax is another important illustration. The separation of the bast fibres of the plant from the cuticular tissues on the one side, and the woody stem on the other, is greatly facilitated by the breaking down of the parenchymatous tissue with which the bast cells are in contact; it is this tissue which rapidly 'rots' under the treatment, and the process is another illustration of 'natural' differentiation of cellulosic tissues.

The third group of decompositions involves the much debated question of the 'fate' of cellulosic tissues in their passage through the alimentary canals of animals, or, to put it more narrowly, the feeding or nutritive value of cellulose. We may take it that the typical cotton cellulose would not be sensibly affected by a passage through the most powerful processes of animal digestion. There are, on the other hand, a number of celluloses which would be, and undoubtedly are, readily digested; and the further consideration of this point may be deferred until we have dealt with the specific differences exhibited by the various members of the cellulose group, more particularly in relation to acid and alkaline hydrolyses, under which groups of decompositions the digestive processes may be generally included.

Special mention may be made of the results of an investigation, by Horace Brown (J. Chem. Soc.), of the question of the presence of a cellulose-dissolving enzyme in the digestive tract of herbivora. After exhaustive inquiry the author establishes the conclusion that the enzyme is secreted by the plants themselves, and comes into activity under the favourable conditions provided by the digestive organs and processes of the animal.

D. DECOMPOSITION BY HEAT.—DESTRUCTIVE DISTILLATION.—Destructive distillations of cellulosic raw materials constitute an extremely important group of industrial processes; and if we include the coals in such a classification, the hydrocarbons of coal tar must be regarded as products of a series of transformations of cellulose, of which the final stages are determined by destructive distillation. By the direct action of heat, however, upon the celluloses proper, 'aromatic' products — hydrocarbons and phenols—are obtained in relatively small quantity. The main products are (1) gases: carbonic anhydride, carbonic oxide, and methane; (2) liquids: water, acetic acid and furfural, methyl alcohol, and small quantities of hydrocarbons and phenols; (3) solids: paraffins and aromatic hydrocarbons in small quantity, and the residual *charcoal*.

The proportions of these products necessarily vary with all the conditions of the distillation, chiefly (1) rapidity of heating and (2) maximum temperature attained. Recent investigations, in which these conditions were carefully regulated and the products of distillation estimated, have led to more definite results than those of previous date. It must be borne in mind that the term Cellulose *has* been used in a somewhat loose way, and by some writers or compilers of articles as synonymous with the cellulosic raw materials generally. 'The destructive distillation of *cellulose*' has in consequence been described as including the woods. It is important now to differentiate between the products obtained from the typical cotton cellulose and compound celluloses, such as the woods. These differences will be noted more particularly when treating of the latter group. At this point we give the results obtained for (*a*) raw cotton, (*b*) bleached cotton (Ramsay and Chorley, J. Soc. Chem. Ind. 11, 872), and (*c*) cellulose (cotton) regenerated from solution as thiocarbonate.

	(a)			(b)		(c)	
	(1)	(2)	(3)	(1)	(2)	(1)	(2)
Weight (grms.)	45	60	50	67	45	54	50
Charcoal, p.ct.	33·33	30·00	33·00	34·33	34·44	36·0	42·0
Distillate ,,	53·33	50·00	46·00	43·32	51·11	43·0	44·0
Carbon dioxide, p.ct.	6·66	9·53	11·00	5·22	7·77	10·0	7·4
Other gases (diff.)	6·68	10·47	10·00	17·13	6·68	11·0	6·6
	100·00	100·00	100·00	100·00	100·00	100·0	100·0

Composition of Distillate.

P.ct. of cellulose							
Acetic acid	—	2·44	1·31	1·75	2·11	1·5	2·0
Methyl spirit	—	—	7·07	3·94	10·24	—	—
Tar	—	8·33	12·00	9·70	13·33	—	—

Gases.

	C.c.	C.c.	C.c.	C.c.	C.c.	C.c.	C.c.
Vol. per 100 grms. excluding CO_2	4,900	4,500	7,000	2,240	2,200	8,000	

Composition p.ct.

Carbon monoxide	76·90	85·74	76·20	54·14	52·46	80·0	
Oxygen	3·66	2·80	3·34	8·50	4·73	4·0	
Residual gas	19·44	11·46	20·46	37·36	43·11	16·0	

The decomposition by heat is accompanied, in the case of two of the above products—viz. the raw cotton (a), and the cellulose regenerated from the thiocarbonate solution (c)—by a well-marked exothermic reaction.

The distillation being carried out in a glass flask heated in an air bath, and the temperatures within the flask and in the surrounding air space being carefully noted, it is observed that at about 325° the former rises suddenly several degrees, and the rise is accompanied by a rush of gases. The reaction is not observed, however, in the case of the bleached cotton.

This exothermic resolution of the molecule we are not yet in a position to interpret, though we may conclude that it is the expression of some special constitutional feature. It is attended with the formation of gaseous products, of which the greater proportion are the oxides of carbon. It will be seen that, although the proportions of gaseous products vary considerably, the ratio of CO_2 to CO shows a general concordance, and is approximately that of their molecular weights. There is, therefore, ground for supposing that the disruption of the molecules is preceded by the accumulation of oxygen in the one direction, and of hydrogen in the other direction, within the molecule, reaching a maximum with the formation of a group $\genfrac{}{}{0pt}{}{CO}{CO}{>}O$, which is then split off explosively, and at the same time resolved. The complementary phenomenon is the further condensation of the residues to form the 'pseudocarbon,' or charcoal, in which the carbon is accumulated relatively to the hydrogen and oxygen, and contains approximately two-thirds of the carbon of the original cellulose.

The constitution of the carbonaceous residues of the process— or charcoals—is at present problematical. The subject has been discussed by the authors, in a paper on the Pseudocarbons (Phil. Mag. May 1882), a name suggested for the designation of this group of compounds—which may be taken to include the coal series. This paper contains a general discussion of the composition of these substances—chiefly devoted to showing that they are not to be regarded as containing 'free' carbon. They are, in fact, $C.H.O$ compounds, and yield derivatives with chlorine, nitric acid, and sulphuric acid, similar to those obtained by Sestini from the bromic or ulmic group of compounds.

Synthesis of Cellulose.—With a large number of carbon compounds it is possible to dissect them molecularly in such a way that the component groups or residues may be put together and the original molecule or compound reconstituted. This is the ordinary history of the synthesis of these compounds, of which the modern science furnishes innumerable instances. In the case of cellulose only one process has been described which may be considered as a constitutional dissection, and that is, the breaking down of the molecule by sulphuric acid. In the final result the process may be interpreted as a simple hydrolysis into dextrose molecules—that is, the acid plays an intermediate part only, combining with the molecule by simple synthesis, and interchanging with water molecules in presence of excess of the latter. The intermediate terms of the dissection process are not sufficiently under control to be followed with that degree of precision which is possible in the case of other complex carbohydrates, notably starch, which are hydrolysed by relatively minute quantities of enzymes or 'unorganised ferments.' Even if this were possible there appears at present no prospect of building up the cellulose molecule by reversal of the process, as our much more complete knowledge of the starch molecule has brought with it no suggestion of a constructive process following inversely the lines of its hydrolytic dissection.

It appears, therefore, on the experimental evidence that cellulose is built up of molecules of simple carbohydrates, but in what manner there are none but hypothetical indications. On the other hand, certain processes have been brought to light which are undoubtedly direct syntheses of cellulose from particular carbohydrates of low molecular weight. Of these two may be cited as typical, one of which (*a*) is due to the action of an unorganised ferment resembling diastase, the other (*b*) is produced by a micro-organism.

(*a*) As a result of a change which is observed to be set up 'spontaneously' in beet juice, a white insoluble substance is formed, and separated in lumps or clots; this substance has all the characteristics of cellulose. After separating this insoluble cellulose the solution gives with alcohol a gelatinous precipitate resembling the hydrates of cellulose previously described. These results are independent of the so-called viscous or mucous fermentations. That the process by which the cellulose is formed has the essential features of a fermentation process, is seen from the fact that when the lumps or clots are transferred to a solution of pure cane sugar, or beet molasses, a further formation of the cellulose ensues. When the process proceeds in neutral solution no carbonic anhydride is evolved; but in presence of acids this gas is evolved, and at the same time acetic acid is formed in the solutions.

E. Durin, by whom these phenomena have been investigated, (Compt. Rend. 82, 1078; 83, 128), regards the ferment as allied to diastase, and states that fresh solutions of diastase itself act on solutions of sugar to form the soluble cellulose, precipitable by alcohol. There is also some evidence that cellulose may be formed from cane sugar in the plant by processes of this kind. It may be noted here that the general view current amongst plant physiologists has been that 'starch is the material from which plants elaborate their tissue substances or cellulose.' The recent researches of Brown and Morris, however, have rather discredited this view, their elaborate and ingenious experiments going to show that cane sugar is probably the immediate mother substance from which the plant cell builds up cellulose, starch being rather a reserve form for what may be regarded as the excessive energy of assimilation in sunlight, being in turn hydrolysed as required to feed the more continuous process of tissue formation.

(*b*) A. J. Brown has more recently made observations upon 'An Acetic Ferment which forms Cellulose' (J. Chem. Soc. 49, 432). The 'vinegar plant' takes a membranous form, which in microscopic examination is seen to be clearly differentiated from the zoogloea form of the Bacterium Aceti. It is, in fact, composed of bacterial rods of 2μ length contained in a membranous envelope. This envelope has the properties and composition of cellulose.

Pure cultures of the organism placed in solutions of levulose, mannitol, and dextrose, reproduce the growth in question, composed, i.e. of the bacteria enveloped in a 'collecting medium' of cellulose. The proportion of cellulose formed, to the soluble carbohydrate disappearing, is highest in the case of levulose.

It is remarkable that the cellulose formed, when hydrolysed by sulphuric acid, gives a dextro-rotary sugar. The organism also has the power of determining the oxidation of ethyl alcohol to acetic acid, and of dextrose to gluconic acid. But its characteristic property is that of the building up of cellulose from the carbohydrates of lowest molecular weight, whence its descriptive name Bacterium Xylinum.

The synthesis of cellulose is a problem involving the whole question of 'assimilation' of 'organic' substance by the plant. It has been held generally by physiologists for a long time that starch is the first visible product of assimilation in the plant cell. On this subject the student should read Sachs's classic work on 'Vegetable Physiology,' the investigations of this observer having contributed in a very important degree to the establishment of the above view. *A priori*, perhaps, it appears somewhat singular that the plant should invariably proceed by way of starch to the elaboration of its permanent tissue. Recent researches of Horace Brown and G. H. Morris (J. Chem. Soc. 1893, 604) throw doubt upon the conclusion from the experimental side. Again, we recommend to the student a careful study of the work of these authors, not merely for the results obtained and described, but for the excellent plan of the investigations. We give a few of the main conclusions in which these investigations issued. 'It is perfectly true, as pointed out by Sachs, that starch is the first *visible* product of assimilation; yet there can be little doubt (as was, in fact, anticipated by Sachs himself) that between the inorganic substances entering into the first chemical process of assimilation and the starch there is a whole series of substances of the sugar class, and that it is from the last members of this series that the chloroplasts, under normal conditions, elaborate their starch. Both under the natural conditions of assimilation and the artificial conditions of nutrition with sugar solutions the chloroplasts form their included starch from antecedent sugar.'

Observations on the secretion of diastase by the leaves of

flowering plants, the variations of diastatic activity with the conditions of assimilation, and the relations of diastase to the starch and sugars (including maltose) present in the leaves lead to the important conclusions which we give in the words of the original :

'Looking at the results all round, they are, it seems to us, decidedly opposed to the view that either dextrose or levulose is the first sugar formed by assimilation, and point to the somewhat unexpected conclusion that, at any rate in the leaves of Tropæolum, *cane sugar* is the first sugar to be assimilated by the assimilatory processes. There seems every reason to believe that this cane sugar . . . functions in the first place as a temporary reserve material, and accumulates in the cell sap of the leaf-parenchyma when the processes of assimilation are proceeding vigorously. When the degree of concentration of the cane sugar in the cell sap and protoplasm exceeds a certain amount, which probably varies with the species of plant, starch commences to be elaborated by the chloroplasts, this starch forming a somewhat more stable and permanent reserve material than the cane sugar, a reserve to be drawn upon when the more easily metabolised cane sugar has been partially used up.'

From these authors' experiments it also appears that, in the translocation of the sugar through the leaf stalk into the stem, it takes the form of dextrose and levulose. The former, however, being more quickly used up in the respiratory process, there is a larger proportion of the latter passing over into the general metabolic circulation.

The starch, on the other hand, migrates in the form of maltose, and this appears to be, in a sense, a starvation phenomenon—that is, it is only put under contribution to the general supply of nutrient material when, and in proportion as, the carbohydrates of lower molecular weight are used up.

These researches obviously constitute an important advance towards the elucidation of the elaborating functions of the plant cell. What the actual first step may be in the building up of tissue-substance, is still a matter of conjecture. The prominent facts presented to us are, (1) that carbonic anhydride is decomposed in the plant cell, the whole of the carbon being retained, and part of the oxygen restored to the atmosphere ; (2) that this decomposition takes place under the influence of the protoplasmic contents of the living cell; but, although, therefore, nitrogen must be regarded as essential to the process, the plant builds up *non-nitrogenous* materials, both immediate and ultimate. (3) That the source of energy which determines these constructive changes is that of the sun's rays; that portion of the solar radiation chiefly concerned being included between the wave lengths $\frac{40}{100000}$—$\frac{78}{100000}$ mm., with a maximum effect corresponding to the yellow-green of the spectrum.

Generally, it may be fairly assumed that the CO_2 of the atmosphere is 'loosely' synthesised with protoplasmic products or chlorophyll,[1] and so brought within the range of the specific molecular activities, representing what we know in the aggregate as vitality.

Constitution of Cellulose.—From the preceding general account of the properties and reactions of the typical cotton cellulose we might be expected to be able to deduce its constitutional formula. We have, however, already pointed out that no purely chemical synthesis of any compound similar to cellulose has been attempted; we are, therefore, without the essential criterion of the correctness of any general formula which might be proposed, if only as a condensed expression of the relationship and functions of its constituent groups.

But although no such formula can be proposed having any but a speculative and a tentative value, it will be a useful guide to future investigation to sum up those reactions which throw a direct light upon the function of the molecule as a whole, and of its constituent groups.

(1) The resolution by sulphuric acid, and subsequent hydrolysis of the esters formed in the reaction, into simple carbohydrate—viz. dextrose molecules. Cellulose is, therefore, in this sense an anhydro-aggregate of the aldose groups $C_6H_{12}O_6$.

(2) Partial resolution under the action of hydrochloric acid, attended by the setting free of CO groups.

In cellulose the carbonyl groups are 'suppressed'; that is,

[1] This view is specifically formulated by E. Fischer, Berl. Ber. 1894., 3231. (Dec. 10).

they either exist in combination—as in the acetals—or are susceptible of an alternative form, the carbonyl becoming hydroxyl oxygen.

(3) Complete proximate resolution, by 'fusion' with alkaline hydrates, into hydrogen, carbonic, oxalic, and acetic acids. The yield of the latter tending to a maximum of 30–35 p.ct. indicates that the grouping $CO—CH_2$ is an important element in the constitution of the unit groups.

(4) *Negative characteristics.*—These are (*a*) those which characterise generally the *saturated* compounds—in which group cellulose must be classified. (*b*) Resistance to alkaline hydrolysis. (*c*) Resistance to oxidising actions up to a certain limit of intensity. (*d*) Resistance to acetylation; requiring either very high temperature or the presence of an auxiliary ($ZnCl_2$) for the determination of reactions of its OH groups with the acid oxide.

(5) *Synthetical reactions.*—Of these the more definite are those which yield the esters, viz. nitrates, acetates, and benzoates. The highest nitrate obtainable appears to be the trinitrate (hexanitrate in the C_{12} formula); the highest acetate the tetracetate (C_6 formula). A higher degree of acetylation has been obtained, but there is undoubted evidence that this results from molecular resolution (hydrolysis). The conclusion to be drawn from the relationship of these esters to the parent molecule is that, of five O atoms in the formula $C_6H_{10}O_5$, four react as OH oxygen with retention of the original configuration of the molecule.

The thiocarbonate reaction further elucidates the functions of the OH groups, and the resistance of the molecule to hydrolysis. It constitutes a further distinction of the celluloses from starch, as a type of molecular configuration; starch failing to give any definite indications of this reaction, and, in contrast to cellulose, being eminently susceptible of hydrolytic resolution.

The Typical Cellulose and the Cellulose Group

To sum up these more prominent points in the evidence of constitution, we are entitled to regard cellulose as conforming, in regard to its ultimate constituent groups, to the general features of the simpler carbohydrates of well-ascertained constitution, but differentiated by a special molecular configuration resulting in a suppression of activity of the constituent groups in certain respects, but on the other hand conferring greater reactivity is others. This molecular configuration involves primarily the question of the mode of arrangement of the carbon with the qualifying hydrogen atoms within the unit groups—which, for the reasons given, may be assumed to be of the dimensions of C_6; and, secondly, the grouping of these into the aggregate which may be regarded as constituting the true molecule of cellulose. Next in importance are those modifications of configuration which are bound up with the disposition of the C atoms.

In regard to carbon configuration the evidences are rather indirect than determinable by the actual properties of cellulose itself. The choice obviously lies between a chain and cyclic formula for the unit groups. The balance of evidence is in favour of the latter and on the following grounds: (1) the general differentiation of cellulose in regard to stability, which points to a symmetrical formula, as distinguished from the normal chain upon which the hexoses are represented; (2) the formation of a cellulose acetate of the composition $C_6H_6O(OAc)_4$, in which only $2n$ carbon valencies are taken up in 'outside' combination; (3) the simple and manifold transitions of cellulose—in the plant world—into keto R. hexene and benzene derivatives. The process of lignification in the plant cell is characterised by the formation of groups of the general form

$$CO\underset{\underset{(OH)_2\ (OH)_2}{C\text{---}C}}{\overset{CH=CH}{\diagup\diagdown}}CH_2$$

which remain intimately associated with the cellulose, of the cell or fibre in combination, as a compound cellulose, therefore (lignocellulose, see p. 137). These derived celluloses exhibit a close general conformity with the parent type—that is, apart from, or in addition to, the special properties and reactions due to the presence of the hexene ring, all the typical characteristics of the cellulose proper.

Although, however, the hexene ring is thus shown to be represented in compounds identified with the 'organic' functions of the plant cell, this does not appear to be the case with the fully 'condensed' benzene ring. Aromatic compounds are formed in profusion, it is true, in the general range of plant life, but when they appear it is in the unorganised form, i.e. as excreted products of metabolism. The same appears also to hold for the terpene series.

It may also be noted here that the supplies of raw materials—hydrocarbons &c.—for the enormous modern industry in 'aromatic' products are derived from the products of coal distillation, and therefore may be traced back to a cellulosic origin.

The Cellulose Group.—Thus far we have been dealing mainly with one member of the very numerous class of plant constituents comprehended in the term 'cellulose.' While the properties and characteristics of cotton cellulose are in suchwise representative that this substance may be regarded as the typical cellulose, the differentiation of this, as of every other group of tissue constituents, in conformity with functional variation, necessarily covers a wide range of divergencies.

The celluloses of the plant world, so far as they have been investigated from the point of view of chemical constitution, group themselves as follows :

(a) Those of maximum resistance to hydrolytic action, and containing no directly active CO groups.

(*b*) Those of lesser resistance to hydrolytic action, and containing active CO groups.

(*c*) Those of low resistance to hydrolysis, i.e. more or less soluble in alkaline solutions and easily resolved by acids, with formation of carbohydrates of low molecular weight.

Group (*a*).—In addition to the typical cotton cellulose — which, it is to be noted, is a seed-hair—there may be included in this group the following fibrous celluloses which constitute the bast of exogenous flowering annuals : viz. the celluloses of *Flax* (Linum usit.), *Hemp* (Cannabis sativa), *China Grass* (Rhea and Boehmeria species) ; and of the lesser known Marsdenia tenacissima, *Calotropis* (gigantea), *Sunn Hemp* (Crotalaria juncea).

As in the case of cotton, the celluloses of the fibres are considered in the form of the white (or bleached) and purified residues resulting from the treatment of the raw materials by processes of alkaline hydrolysis and oxidation more or less severe in character. For the purification of the celluloses in the laboratory the methods usually practised consist in (1) alkaline hydrolysis, i.e. treatment with boiling solutions of sodium hydrate, carbonate, or sulphite ; (2) exposure to bromine water or chlorine gas ; or when oxidation alone is sufficient for the removal of the 'impurities,' to solutions of the hypochlorites or permanganates (in the latter case followed by a treatment with sulphurous acid to remove the MnO_2 deposited on the fibre-substance) ; (3) repetition of (1) for the removal of products rendered soluble by (2).

Special accounts of these raw fibrous materials are contained in Spon's 'Encyclopædia Industrial Arts' ; 'Die Pflanzenfaser,' Hugo Müller (A. W. Hofmann's 'Bericht.' Braunschweig, 1877) ; 'Report on Indian Fibres and Fibrous Substances,' Cross, Bevan, and King (Spon, London, 1887) ; and 'Chemische Technologie d. Gespinnstfasern,' O. N. Witt (Braunschweig, 1888).

It has been already pointed out that these celluloses occur in admixture or combination with other substances, often grouped together in the term *non-cellulose*; cellulose and non-cellulose being usually separated jointly from the plant, and constituting the 'raw fibre.' The raw fibre is therefore usually a *compound cellulose*, though in some cases a compound of a very weak order. These points will be best illustrated by a careful study of commercial flax. Flax is made up of the pure fibre, which is a compound cellulose, with a certain admixture of the tissues with which it is in contact in the stem. These adventitious components are largely got rid of, first in the processes of breaking and scutching, and afterwards in the further refining processes of hackling and preparing, by which the spinner brings the fibre into the proper condition for the twisting or spinning process proper. But the yarn still retains residues of the cuticular cells and wood (sprit), which then require to be broken down, or converted into cellulose, by the chemical processes of bleaching. It is the former which occasion the major difficulties of the linen bleacher. As a result of the intimate association of the fibre with the cuticle of the stem, flax, as finished for the market, contains an unusually large proportion of oil-wax constituents, i.e. from 3–5 p.ct. of such bodies, soluble in the special solvents. These may be separated by fractionation into (*a*) ceryl alcohol and derivatives (esters), and (*b*) a mixture of oily bodies of ketonic character.

For more detailed investigation of this group of flax constituents see Hodges, Proc. R. I. Acad. 3, 460; and Cross and Bevan, J. Chem. Soc. 57, 196.

This oil-wax complex plays an important part in the ordinary process of flax 'line' spinning, and the failure of many of the artificial processes of 'retting' flax may be explained by the deficiency of the resulting fibre in these constituents. In the breaking down of the cuticular celluloses, whether in the retting (rot-steep) or bleaching process, these waxes and oils are separated. Their elimination from the cloth necessitates the very elaborate treatment by which the 'Belfast Linen Bleach' is obtained.

These constituents are adventitious impurities, the bast fibre itself being a pectocellulose (see p. 214), easily resolved by alkaline saponification into cellulose on the one hand, and soluble modification of the pectic group on the other. Although, therefore, the

The Typical Cellulose and the Cellulose Group

considerable loss of weight of flax cloth in bleaching (20-30 p.ct.) falls mainly in the early alkaline treatment, the chief difficulties are in the breaking down of the more resistant bodies derived from the cuticle, including chlorophyll.

The celluloses of this group thus purified may be taken as chemically identical with cotton cellulose, investigation having so far failed to differentiate them. It must be noted, however, that the several members of the group present distinct morphological characteristics, and differ also in such external properties as lustre and 'feel.' These are in part correlated with the differences in minute structure, but they are no doubt in part differences of substance. So far, however, we have no knowledge of the proximate constitution of these substances, and can therefore say nothing as to the causes of difference in this respect.

On the other hand, the essential identity of these celluloses is established in regard to ultimate composition and in reference to the following properties and reactions :

(1) Resistance to hydrolysis and oxidation, and other negative characteristics, indicating a low reactivity of the CO and OH groups.

(2) Their relationships to the special solvents previously described, including the thiocarbonate reaction.

(3) Formation of esters, nitrates, acetates, benzoates.

Of the above, it is sufficient in general laboratory practice to examine cellulose in regard to ultimate composition, resistance to alkaline hydrolysis, behaviour with solvents, and reactions with sulphuric acid (solution without blackening) and nitrating mixture (H_2SO_4 and HNO_3); the 'nitration' proceeds without oxidation, and gives a higher yield of product, 160-180 p.ct. according to the condition.

Group (b).—These celluloses are differentiated from the former group (1) by ultimate composition, the proportion of oxygen being higher ; (2) by the presence of active CO groups ; (3) in certain cases by the presence of the $O.CH_3$ group.

The general characteristics of the group are those of the *oxycelluloses*. It has recently been shown that these oxidised derivatives of the normal celluloses are further characterised by yielding *furfural* as a product of acid (HCl) hydrolysis. The yield of this aldehyde is, in certain cases, increased by previous treatment of the oxycellulose with a reagent prepared by saturating sulphuric acid of 1·55 sp.gr. with HCl gas. In this reagent the oxycelluloses dissolve; and on then diluting with HCl of 1·06 sp.gr. and distilling, maximum yields of furfural are obtained, the yield being a measure of the increased proportion of oxygen beyond that corresponding with the formula $C_6H_{10}O_5$.

Celluloses of this class are much more widely distributed in the plant world than those of the cotton type ; they appear, from recent observations, to constitute the main mass of the fundamental tissue of flowering plants, in which they usually exist in intimate mixture or combination with other groups more or less allied in general characteristics. It appears, from a survey of the contributions of investigators to the subject of cellulose, that research has been very much confined to the fibrous celluloses, more particularly to such as receive extended industrial use. The time has come, however, when systematic research is much needed to establish at least a preliminary classification of the 'cellular' celluloses upon the lines of chemical constitution. Constitution, taken in relation to physiological function, is an attractive subject of research ; and it is in the plant cell, where synthetical operations are predominant, that we have to look for the foundations of the 'new chemistry,' which may be expressed broadly as the relation of matter to life.

It is to be noted that the differentiation of many of these celluloses from the typical cotton is, in regard to empirical composition, only slight. There appear, on the other hand, to be more important differences of constitution. Thus pine-wood cellulose dissolved in sulphuric acid, the solution diluted and boiled, and further treated by the isolation of crystallisable carbohydrates, yields these (i.e.

dextrose) in only small proportions. (Lindsey and Tollens, Lieb. Ann. 267, 370.)

Investigation has stopped short at this negative result. It would be of interest, therefore, to isolate the products formed in the reaction with the concentrated sulphuric acid, so as to characterise them, at least generally. Until this is done, or some other method proximate resolution is worked out in detail, we can only say that the constitution of these celluloses is in some important feature radically different from that of the typical cellulose.

An account of recent investigations of these 'celluloses' will be found in Berl. Ber. 1893, and a more special treatment of the subject, *ibid.* 1894, and J. Chem. Soc. 1894 (C. Smith).

Of this group of the natural oxycelluloses the following have been more particularly investigated :

(1) *Celluloses from woods and lignified tissues generally.*—Lignified tissues are made up of compound celluloses, to be subsequently described (see Lignocelluloses, p. 91), from which the celluloses may be isolated by a number of treatments, all depending upon the relative reactivity of the so-called 'non-cellulose' constituents, which in combination with the celluloses make up the compound cellulose, lignocellulose or wood substance. These non-cellulose constituents are readily attacked and converted into soluble derivatives; and there are various industrial processes for preparing celluloses (paper pulp) from raw materials of this class, depending upon the direct conversion of the former into such soluble compounds. The isolated celluloses show the following general characteristics (Berl. Ber. 27, 1061) :

Elementary composition $\begin{cases} C & 42\cdot8-43\cdot8 \text{ p.ct.} \\ H & 5\cdot6-5\cdot9 \text{ ,,} \end{cases}$ *Yield of furfural*, by solution and hydrolysis (HCl), 2-6 p.ct. *Reactions* with phenylhydrazine salts and magenta-sulphurous acid, indicating the presence of active CO groups. These celluloses are necessarily less resistant to oxidation and hydrolysis, but show in all other respects a close general agreement with the normal cotton cellulose.

(2) *Celluloses from cereal straws, from esparto, &c.*—These celluloses are isolated from the matured stem, or haulm, by digestion with alkaline lye at elevated temperatures. They are also of considerable industrial importance, being largely used in the manufacture of the cheaper kinds of writing and printing papers.

Recent investigation has shown that these celluloses are strongly differentiated from the normal, and are in fact pronounced oxycelluloses. The following are the characteristics of difference :

Ultimate composition, after treatment with hydrofluoric acid to remove siliceous ash constituents :

	Oat straw cellulose		Esparto cellulose	
	(1)	(2)	(1)	(2)
C	42·4	42·4	41·78	41·02
H	5·8	5·8	5·42	5·82

Yield of furfural by solution and hydrolysis (HCl) :

Oat straw cellulose	Esparto cellulose
12·5	12·2

Reactions.—In addition to those with Fehling's solution, phenylhydrazine salts, and magenta-sulphurous acid indicating the presence of active CO groups, the celluloses give a characteristic rose-red colouration on boiling with solutions of aniline salts. This reaction serves to identify their presence in papers, and from the depth of the colouration, the percentage may be approximately estimated.

Investigation has also established the following points in regard to the oxidation and deoxidation of these oxycelluloses.

They are gradually oxidised in dry air at the temperature of the water-oven, undergoing discolouration ; the yield of furfural, by hydrolysis, showing a progressive increase. They are deoxidised, on the other hand, by neutral and alkaline reducing agents. Thus after lengthened exposure to solutions of zinc-sodium hyposulphite, prepared by the action of zinc dust upon sodium bisulphite, the yield of furfural—which is a measure of the degree of oxidation—was reduced, in the case of esparto cellulose, from 12·6 to 8–9 p.ct.

A still further deoxidation results from solution of these oxycelluloses as thiocarbonate, and regeneration of the cellulose by heating the solution at 80–100°. The regenerated cellulose approximates to the normal, yielding only 2 p.ct. furfural on hydrolysis. It is to be noted, however, that esparto cellulose, in common with all the celluloses of this group, is partly hydrolysed to soluble derivatives by this treatment; the regenerated cellulose amounting to 80 p.ct. of the original weight dissolved. The soluble portions yield furfural on hydrolysis, amounting (in a typical experiment) to 4·0 p.ct. of the original.

The celluloses of this group are dissolved by concentrated sulphuric acid to dark coloured solutions. On diluting and boiling they are resolved into carbohydrates of low molecular weight; dextrose appears to be invariably formed, and in many cases also mannose; but only very small yields of either carbohydrate have been so far obtained.

Group (c).—This includes the heterogeneous class of non-fibrous celluloses which we have defined as of low resistance to hydrolysis, being easily resolved by boiling with dilute acids, and being also more or less soluble in dilute alkaline solutions. This group has been but little studied, and therefore can only be generally characterised. Physiological research has shown that there are a large number of cellular, as distinguished from fibrous 'celluloses,' which are readily broken down (hydrolysed) by the action of enzymes within the plant itself, whether as a normal or abnormal incident of growth. Thus in the germination of starchy seeds, the cell walls (cellulose) of the starch-containing cells are broken down, as a preliminary to the attack upon the starch granules themselves, to form the supply of nutrition to the embryo. In an exhaustive investigation of the germination of the barley, Brown and Morris have thrown a good deal of light upon this particular point, which they emphasise in the following words: 'that the dissolution of the cell wall invariably *precedes* that of the cell contents during the breaking down of the endosperm is a fact of the highest physiological importance, and one which for the most part has been strangely overlooked.'

A similar, but abnormal, dissolution of cell walls is that which occurs in the attacks of parasitic organisms upon the tissues which they invade.

These processes are well known to physiologists, who, however, generally regard 'cell-wall' and 'cellulose' as substantially identical terms. The chemical differentiation of the substances comprising cell walls is, on the other hand, an entirely new field of research; but although investigation has not gone very far, the results are sufficient to show that the celluloses of this order are enormously diversified. The variations already disclosed are (1) those of the carbohydrates yielded by ultimate hydrolysis, and (2) those of molecular configuration or condensation. We have already seen that the celluloses of the cotton group (*a*) yield dextrose as the ultimate product of hydrolysis; those of group (*b*) yield, in addition to dextrose, mannose and probably other bodies; and the group we are at present discussing yield, in addition, galactose, and the pentoses xylose and arabinose. In illustration we may cite a few examples. Thus GALACTOSE has been obtained as a product of hydrolysis of the cell walls of the seeds of Lupinus luteus, Soja hispida, Coffea arabica, Pisum sativum, Cocos nucifera, Phœnix dactylifera, &c. MANNOSE is obtained in relatively large quantity from the 'ivory nut,' and from a very large number of other seeds; and PENTOSES, from the seeds of the cereals and of leguminous plants. It appears, therefore, generally that a large number of plant constituents which have been denominated by the physiologists as 'Cellulose' have little more title to be considered as such than has starch. However, external resemblances count for something, at least in the beginnings of classification, and substances of the type we are considering may be conveniently grouped with the celluloses; but we should propose to apply to them the term PSEUDO-CELLULOSES, or HEMICELLULOSES—as has been proposed by E. Schulze. Our group (c) of pseudo-celluloses may therefore be defined as—Substances closely resembling in appearance the true celluloses, but easily resolved into simple carbohydrates by the hydrolytic action of enzymes, or of the dilute acids and alkalis.

Animal Celluloses.—*Tunicin*—a compound of the empirical composition of cotton cellulose, and resembling it in a number of its properties and reactions—is isolated from the mantle of Ascidia and other invertebrate species by exhaustive hydrolytic treatments. Such resistant residues have been investigated by Schmidt (Annalen, 54, 318), Berthelot (Compt. Rend. 47, 227), Löwig and Kölliker (J. Pr. Chem. 39, 439), Schäfer (Annalen, 160, 312), and more recently by Franchimont (Compt. Rend. 89, 755). From these later investigations it appears that the sugar obtained as the product of ultimate hydrolysis is identical with the dextrose obtained from the vegetable celluloses. From this and its reactions generally, which differ in some respects from those of the normal cellulose, Franchimont concludes that the compound is undoubtedly a cellulose, but of different constitution from the normal. Cellulose has also been identified as a constituent of the protozoa. Investigations of one of these organisms—*Ophrydium versatile*—by Halliburton showed the investing matrix of a colony of these ciliated protozoa to consist in the main of a cellulose similar to that of the Tunicata (Q. J. Micr. Soc. 1885, 445).

These scattered observations indicate that the special constitutional type or configuration of the celluloses is not confined to those of vegetable origin. There is, of course, no reason in the nature of things that the distribution of the type should be limited to the plant world. It is quite possible, in fact, that the animal fibres, and more generally the colloids of the animal skeleton, may prove to be of similar carbon configuration to that of the celluloses. A systematic investigation of such a possibility has, so far as we know, not been attempted. Suggestions have been made—in, it is true, rather a wild way—that the silkworm is engaged in converting the cellulose of the mulberry leaf into silk. It is impossible to say, *à priori*, how far the digestive processes in an organism of this order may be destructive in character, but an exhaustive physiological investigation would throw light on the point. It is clear, of course, that the animal organism is not constructive in the same sense as the plant cell, and it is an interesting subject for speculation and experimental inquiry how far the vegetable products constituting the food of animals are broken down by the digestive process; or, in other words, how far they may preserve their constitutional features in being synthesised to 'animal' products.

PART II

COMPOUND CELLULOSES

IN dealing with the isolated celluloses it has been shown that the processes by which they are isolated or purified are based upon the relative reactivity of the compounds with which the celluloses are combined or mixed, in the raw or natural products of plant life. These natural forms of cellulose are, of course, multitudinous. Remembering the infinite variety of the vegetable world, the endless differentiation of form and substance of the tissues of plants, it might be presumed that the chemical classification of these products would present unusual complications.

Investigation, however, has shown, and continues to show, that this great diversity of substance, as revealed by proximate analysis, exists upon a relatively simple chemical basis. The compounds constituting the fundamental tissue of plants may, in fact, be broadly classified in correspondence with the three main types of differentiation of the cell wall long recognised by the physiologists, viz. *lignification, suberisation*, and conversion into *mucilage*. That is to say, in addition to the celluloses proper and hemi- or pseudo-celluloses—which may be defined as polyanhydrides of the normal carbohydrates, ketoses and aldoses—there are three main types of *compound* celluloses in which the celluloses as thus defined exist in combination with other groups, as follows :

Lignocelluloses.—The substance of lignified cells and

fibres, notably the woods—of which the characteristic *non-cellulose* constituent is a R. hexene derivative.

Pectocelluloses and Mucocelluloses.—Comprising a wide range of tissue constituents—of which the non-cellulose constituents are colloidal forms of the carbohydrates, or closely allied derivatives, easily converted by hydrolytic treatments into soluble derivatives of lower molecular weight, and belonging to the series of 'pectic' compounds, or hexoses, &c.

Adipocelluloses and Cutocelluloses.—The substance of cuticular and suberised tissues—in which the cellulose is associated with fatty and waxy bodies of high molecular weight.

To deal with these groups in detail would involve a survey of the entire vegetable kingdom; of which, on the other hand, but a very small section has been subjected to systematic investigation. It is true, of course, that an immense number of proximate analyses of vegetable products have been put upon record; but the analytical methods adopted have been of the empirical order, and their results, stated under such terms as 'crude fibre,' 'non-nitrogenous extractive matters,' cannot be regarded as 'systematic' in the sense of constitutional diagnosis. We shall therefore confine ourselves to an account of typical members of the above groups, and such as have been investigated by molecular, as opposed to statistical methods.

Frémy has devised (Compt. Rend. 83, 1136) a system of chemical differentiation and classification of vegetable tissue constituents, which, although it has found but little favour, and is in fact generally rejected by critical writers on this subject, may be briefly noted here.

The classification embraces (*a*) celluloses, including 'paracellulose' and 'metacellulose' in addition to the normal cellulose; (*b*) vasculose; (*c*) cutose; (*d*) pectose and pectic compounds; and (*e*) nitrogenous bodies.

The celluloses (*a*) are differentiated by treatment with the cuprammonium reagent: 'cellulose' dissolves directly; 'paracellulose' (epidermis of leaves, &c.) dissolves after boiling with hydrolysing acids; 'metacellulose' (found chiefly in lichens) remains insoluble after the acid treatment.

Vasculose is insoluble in cuprammonium; it is readily dissolved on heating at high temperatures with solutions of the alkalis. It is attacked by all oxidising agents. It may be selectively attacked by dilute nitric acid. Vasculose is said to abound in hard woods, the hard concretions of pears, &c. It appears to be identical with the *lignocellulose* of this treatise; the non-solubility in cuprammonium being a statement of doubtful value.

Cutose is the substance of the transparent cuticular membrane of leaves &c. It has been further studied by Frémy, and the results of the later investigations are given in this treatise, p. 229.

Pectose and pectic constituents.—These have also been further investigated by Frémy, and his results are noted in connection with the pectocelluloses, p. 216.

In the main, therefore, the lines of this classification are adopted in this treatise, but that is probably because, and in so far as, they have a physiological basis. Chemically speaking, the classification is of little value, since it rests chiefly upon the actions of hydrolytic agents. Frémy's experimental work, on the other hand, is of a certain empirical value apart from the conclusions drawn from the results; but as it does not contribute to the solution of constitutional problems it will not be reproduced here. An exhaustive account of the researches will be found in Ann. Agron. 9, 529.

Of the three groups of compound celluloses, the lignocelluloses stand first in order of importance. Not only are they by far the most widely distributed, but they have a physiological and a special chemical significance which mark them out as the arena of some of the most interesting processes presented by the many-sided synthetical activity of the plant cell.

Of the lignocelluloses there are two well-defined types: (1) the bast fibre of the Corchorus species, known in commerce as *jute*; (2) the woods, i.e. the lignified tissues of perennial stems. The former, being a simple tissue and an annual growth, is a more promising subject for the investigation of the general chemistry of *lignification* than the woods, which are, of course, complex structures and subject to continuous modification with lapse of time and in adaptation to the varying necessities of the plant. For this, amongst other reasons, the jute fibre has been more thoroughly investigated than the woods. The results of these investigations will therefore be reproduced at some length. It will simplify the treatment of the subject if we first give a brief account of the fibre-substance in theoretical terms; afterwards the methods of investigation by which the theoretical conclusions have been established will be given in greater detail, and in strict sequence of the lines upon which the celluloses proper have been described.

Lignocelluloses.—(1) **The Jute Fibre.**—The jute fibre-substance differs strikingly in composition and reactions from the celluloses. With its higher $\frac{C,H}{O}$ ratio, viz. $\frac{C\ 46\text{-}47;\ H\ 6\cdot1\text{-}5\cdot8}{O\ 47\cdot9\text{-}47\cdot2}$, there are associated the characteristics of an unsaturated compound—i.e. it contains C=C groupings, and these are localised in C_6 rings. These rings are, further, of ketonic or quinonic character (containing a CO group), and appear to be linked, by O, into complexes of the magnitude of C_{18}. They combine readily with chlorine, in presence of water, and the resulting quinone chlorides are bodies of definite properties and reactions.

A second characteristic constituent of the fibre-substance is a furfural-yielding complex, which appears to be an oxycellulose derivative, a polyanhydride passing by hydration into an oxycellulose of the ordinary type.

The third main constituent is the cellulose of the fibre, which can only be isolated by chemical treatments selectively attacking the 'non-cellulose' in which the two previously described constituents are comprised. The reagents available for the purpose are chiefly the halogens, the halogenated derivatives of the non-cellulose being dissolved away by treatment of the product with alkaline solutions. The cellulose thus isolated is not homogeneous, but is made up of a more resistant α- and a less resistant β-cellulose—more or less resistant, i.e. to the action of oxidants and hydrolytic agents. By other reactions, therefore, in which the oxidising or hydrolysing conditions are more severe, the β-cellulose is converted into soluble derivatives. Such are, digestions with dilute nitric acid; with permanganates, in presence of alkali in excess; with solutions of the bisulphites at elevated temperatures. This β-cellulose is characterised by the presence of $O.CH_3$ groups. The α-'cellulose' is an oxycellulose.

There are other minor characteristics of the non-cellulose portion of the fibre-substance which remain to be noticed. These are (1) the presence of OCH_3 groups, in larger proportion than in the β-cellulose; (2) the presence of a $CH_2.CO$ residue, which is split off as acetic acid, under various hydrolytic treatments of the fibre-substance, probably in union therefore, as a side chain, with the R. hexene groups; (3) the presence of a body giving the characteristic reactions of the pentaglucoses. The pentosans are, in fact, obtainable in small quantity as products of alkaline hydrolysis of the fibre-substance; and the furfural-yielding constituent of the non-cellulose, already described as a condensed oxycellulose derivative, might be assumed, on this evidence, to possess the pentose configuration; but the evidence available so far is not such as to give a definite solution of this point.

These are the main points of constitutional differentiation of the lignocelluloses from the celluloses proper. It has been largely the custom to describe the compound celluloses of this class as *mixtures* of cellulose and non-cellulose, the latter being described generally as 'encrusting matters,' or under the more special term *lignin*, in recognition of its, or their, well-defined individuality.

This view will be found inconsistent with the results of the systematic investigation of this particular fibre-substance, as indeed of the 'lignified celluloses' generally. They are found to be very uniform in composition; the cellulose and non-cellulose are in intimate combination, resisting severe hydrolytic treatment; and in a large number of reactions the typical characteristics of the celluloses are preserved. Therefore the substantive term Cellulose is used in describing them, with the addition of the adjective or qualifying prefix. Where we have to speak of the non-cellulose complex we shall use the term *lignone*, indicating thereby its ketonic characteristics.

We may sum up these outlines of the constitutional features of the jute fibre-substance in a general diagram:

Lignocellulose.

Cellulose		*Lignone* (non-cellulose)	
Cellulose α	*Cellulose* β	*Furfural-yielding*	*Keto R. hexene*
Containing	Containing	*complex*	*group*
oxidised groups	$O.CH_3$ groups	and secondary constituents	
		$O.CH_3$ groups	and	$CH_2.CO$ residue

With this preliminary general survey in view, the experimental treatment of the subject matter will be more readily appreciated in its bearings upon the constitutional problem.

Methods of Quantitative Estimation of Constituent Groups.—The groups above described may be regarded as the proximate constituents of the fibre-substances, and they may be quantitatively estimated by particular methods of proximate resolution, which must be described in some detail.

(1) CELLULOSE.—*Chlorination method.*—For the elimination of the non-cellulose, by conversion into soluble derivatives, various methods are available. One method only gives maxi-

mum yields of cellulose, and for the reason that it is based upon a well-defined reaction of the lignone group, admitting of perfect control. This group, or rather its R. hexene constituent, reacts with chlorine gas, in presence of water, to form a quinone chloride, without, at the same time, affecting its union with the furfural-yielding complex. On treating the chlorinated fibre with sodium sulphite solution, the lignone chloride is dissolved, and at the same time converted into a brilliant magenta colouring matter. The undissolved residue (75-80 p.ct.) is the cellulose of the fibre. The process is carried out as follows :

About 5 grms. of the fibre—weighed after drying at 100°—are (*a*) boiled for 30 minutes with a dilute solution of sodium hydrate (1 p.ct. NaOH), which is kept at constant volume by addition of water. The fibre is well washed on a cloth or wire gauze filter, squeezed to remove excess of water, opened out, placed in a beaker, into which (*b*) a slow stream of washed chlorine gas is passed. Rapid reaction ensues, and the fibre changes in colour, from brown to a bright golden yellow. To ensure complete conversion of the lignone, it is necessary to leave the fibre for some time (from 30-60 minutes) in the atmosphere of Cl gas. (*c*) The chlorinated fibre is removed, washed once or twice with water to remove hydrochloric acid, and placed in a 2 p.ct. solution of sodium sulphite ; the solution is gradually raised to the boiling point, a small quantity of caustic soda solution is added (0·2 p.ct. NaOH calculated on the solution), and the boiling continued for 5 minutes. (*d*) The cellulose is now thrown upon a cloth filter and washed with hot water. It will be found to be almost pure, i.e. white ; but to remove the last residues of the non-cellulose, it may be bleached by immersion in a dilute solution of hypochlorite (0·1 p.ct. NaOCl) for a few minutes, or treated with dilute permanganate solution (0·1 p.ct. $KMnO_4$). It is well washed from these oxidising solutions, treated with sulphurous acid on the filter, well washed with water, squeezed, dried and weighed.

The cellulose estimations by this method give what may be considered the maximum yield; other methods attack the β-cellulose more or less, giving products which are dissolved and removed. These methods may be briefly noticed.

(2) *Bromine water* (Hugo Müller).—This halogen, and in the form of aqueous solution, fails to saturate the R. hexene groups in one operation; hence the alternate treatment with bromine water in the cold, and boiling alkaline solution, requires to be once, twice, or even three times repeated. The yield of cellulose is 2-5 p.ct. lower, and the process is by comparison tedious.

The difference in yield is due to the attendant oxidation and hydrolysis of the β-cellulose.

The following experimental determinations bear upon this point:

A specimen of jute gave the following percentages of cellulose under various methods of treatment:

(1) 73-74 p.ct. Bromine water (cold) and boiling aqueous ammonia alternately till pure.

(2) 74-76 p.ct. Chlorination at ordinary temperatures, followed by alkaline hydrolysis.

(3) 80·9, 80·6, 79·9, 82·0, 81·3, 84·5, in individual experiments in which the treatments were varied as follows: (*a*) chlorination at 0 5°; (*b*) followed by digestion in dilute sulphurous acid at 0 5°; (*c*) hydrolysis with sodium sulphite solution, at first cold, afterwards raised to boiling.

From these results it appears that, by excluding oxidising conditions and graduating the hydrolysis of the chlorinated derivative, a considerable proportion is separated in the form of fibrous cellulose; this portion, under other conditions, is hydrolysed to soluble derivatives.

(3) *Nitric acid and potassium chlorate* (Schulze).—This method consists in a prolonged digestion 10-14 days) of

the fibre-substance, at ordinary temperatures, with nitric acid of 1·10 sp.gr., containing potassium chlorate (0·5–0·8 p.ct. of the weight of the fibre) previously dissolved in the acid. The lignone is attacked jointly by nitrogen and chlorine oxides, and largely converted into derivatives, soluble in the acid solution. The β-cellulose is also considerably attacked (oxidised), and the action extends to the more resistant (α) cellulose. The residues of the lignone are dissolved away by boiling the washed fibre with dilute ammonia.

The process has been largely used in investigations of the lignocelluloses; but the results, both as to yield and composition of the cellulose, are, for obvious reasons, of subordinate value.

(4) *Dilute nitric acid at 50–80°.*—By digesting the fibre-substance with nitric acid (5–10 p.ct. HNO_3) at 60°, the lignone is entirely converted into soluble derivatives; the β-cellulose is also hydrolysed and dissolved; the residue of the treatment being the more resistant α-cellulose. The interaction of the lignone and the acid is of use in elucidating the constitution of the non-cellulose groups, and will be subsequently described from this point of view. As a process of cellulose isolation, the reaction is carried out as follows:

The weighed fibre is placed in a flask and covered with three times its weight of 10 p.ct. HNO_3. It is heated for some hours at 60° until the fibre has changed to a pale lemon yellow colour, the solution being of a bright yellow. After washing away the acid by-products the residual cellulose is boiled with a solution of sodium sulphite, which removes the last traces of lignone derivatives. It may then be finally washed on a cloth filter, squeezed, dried and weighed. The yields of cellulose by this method are 63–65 p.ct. of the fibre.

(5) *Sulphite and bisulphite processes.*—By digestion with solutions of the sulphites of the alkalis, or of the bisulphites

of the alkaline earths, at high temperatures, the lignone groups are attacked and dissolved, as the result of a specific reaction, which will be subsequently dealt with in its bearings upon the constitution of the lignone molecule.

As a 'cellulose process,' the reaction may be carried out as follows :

Neutral sulphite.—The fibre is sealed up with five times its weight of a 6 p.ct. solution of sodium sulphite ($Na_2SO_3.7H_2O$).

Bisulphite of lime or magnesia.—The fibre is sealed up with five times its weight of the bisulphite solution containing 3 p.ct. total SO_2.

The digestion is carried out at high temperatures—either in glass tubes or autoclaves of metal, according to the circumstances of the laboratory. In the latter case, iron vessels may be used with the neutral sulphite, but to prevent reaction with the metal, the solution should contain sodium carbonate ($\frac{1}{6}$ the weight of the sulphite). For the bisulphite treatment a lead-lined digester is required. The maximum temperatures necessary are 180° for the neutral sulphite, 160° for the bisulphite process.

The temperature is raised gradually to the maximum, at which it is maintained for 2–3 hours, the entire duration of digestion necessary being from 6-8 hours. At the expiration of this time, the vessels are cooled off, the contents thrown on to a cloth filter, the residual cellulose washed thoroughly with hot water, and finally purified by treatment with dilute hypochlorite (0·5 p.ct. NaOCl) or permanganate solution. After washing from the oxidising solution, the cellulose is treated with sulphurous acid, from which it is thoroughly washed, squeezed, and dried for weighing.

The yields of cellulose are from 60-65 p.ct. ; the β-cellulose being, under this treatment, also hydrolysed and dissolved.

Furfural-yielding Complex.[1]—It has been found, so far, impossible to isolate this constituent of the fibre-substance; in the mean time, we are limited to indirect methods of arriving at its constitution and its quantitative relationship to the ligno-cellulose molecule. It has been established, by the elaborate researches of Tollens and his pupils, that the condensation to furfural of those carbohydrates which, from their constitution or configuration, readily yield this special product of dehydration, admits of such control that the yield of the aldehyde may be regarded as constant, and an exact measure, therefore, of the parent molecule. The general method of conversion is that of boiling the substance with hydrochloric acid of 1·06 sp.gr. (12 p.ct. HCl). For the estimation of the resulting furfural, various methods have been proposed and practised; the final selection resting with that which consists in converting the aldehyde into its hydrazone, which is then gravimetrically estimated.

A careful survey of the evidence upon which this selection is grounded, together with an elaborate account in detail of the methods both of the conversion of the carbohydrate into the aldehyde in question, and its estimation as described, will be found in a recent paper by Flint and Tollens (Landw. Vers.-Stat. 42, 381–407), which should be closely studied.

The process may be outlined as follows : (*a*) The weighed fibre (5 grms.) is placed in a flask, covered with 100 c.c. of hydrochloric acid of 1·06 sp.gr. The flask is fitted with a double-bored indiarubber cork, carrying (1) the connection to the condenser, the usual bent glass tube, (2) the tubulus of a stoppered 'separating funnel.' The flask is heated, preferably in a bath of oil or fusible metal, so that the rate of distillation is about 2 c.c. per minute. The distillate is col-

---

The designation 'furfural-yielding complex' or carbohydrate may be conveniently shortened to *furfurose* or *furfurosan* in accordance with the modern nomenclature of the group.

lected in portions of about 30 c.c., and when each such quantity is obtained, a corresponding quantity of the acid is admitted to the flask through the separating funnel. The distillation is continued until a drop of the distillate ceases to give the well-known reaction of the aldehyde (rose-red colouration with aniline acetate, in presence of acetic acid).

(*b*) *Conversion into hydrazone.*—To ensure constant results it is important that constant conditions be adopted. The hydrochloric acid is neutralised with sodium carbonate, a slight excess being used; the solution is then made acid with acetic acid. The distillate is made up to a constant volume, and, as it is necessary to keep the proportion of sodium chloride approximately constant, any deficiency of volume is made up with a salt solution of corresponding concentration.

The phenylhydrazine solution is made up with 12 grms. of the base, 7·5 grms. glacial acetic acid, and water to 100° c.c.

The formation of the hydrazone takes place according to the reaction

$$C_5H_4O_2 + Ph.N_2H_3 = C_5H_4ON_2HPh + H_2O.$$

and in any series of determinations with the same substance, the quantity of phenylhydrazine solution necessary to be added is, therefore, approximately known. The quantity is controlled by testing the solution with aniline acetate, drops of the solution being placed upon filter paper moistened with the reagent. The solution is set aside for the separation of the hydrazone, which is much facilitated by continuous stirring. The hydrazone is collected in a filtering tube, containing a perforated plate of platinum or porcelain upon which a circle of filtering paper is laid, the filtration and washing of the precipitate being expedited by the use of the pump. The filter tube, with its contents, is dried preferably *in vacuo* at 60-70°, or in a slow current of dried air at this temperature. It is then weighed. Having been weighed together with the filter paper before the experiment, the difference of the weight gives the weight of hydrazone obtained. From this weight, that of the furfural is calculated as under.

Furfural = [Hydrazone × 0·538]. The factor 0·538 is the mean number obtained from an extended series of experimental determinations. The variations of the results obtained from the theoretical are due to the slight and varying solubilities of the hydrazone in salt solutions of varying concentration. For more exact approximation, the factor must be selected according to the exactly ascertained conditions of its precipitation, and the corresponding value determined by Flint and Tollens (*loc. cit.*) being used in calculating from the hydrazone to the aldehyde.

In the case of the lignocelluloses we are not able to calculate the results to their final expression, i.e. the weight of the parent substance from which the furfural is obtained, for the reason that the constitution of the latter has not been definitely ascertained. The fibre-substance certainly contains pentosanes, and the pentose xylose has been isolated in small quantity as a product of alkaline and acid hydrolysis (Tollens, Berl. Ber. 1889, 1046). As, however, the evidence goes to show that other furfural-yielding groups—probably oxidised hexose derivatives—are present, we are limited to an approximate estimate of the quantity of the entire furfural-yielding complex. This approximation is furnished by multiplying the weight of hydrazone by 1·1 ; or, in other words, the weight of the furfural-yielding constituents may be taken together at about twice the weight of the furfural obtained.

Apart from all hypothetical considerations, however, the yield of furfural is an important constant of the fibre-substance, which may be determined as described, within the limits of satisfactorily small errors of experiment.

Keto R. Hexene Constituent.—The characteristic

reaction of this group is its combination with chlorine, the *quantitative* features of which have been the subject of careful investigation. The chlorinated lignone is a body of definite and uniform composition, represented by the empirical formula $C_{19}H_{18}Cl_4O_9$. It is still a complex containing a quinone chloride, allied to mairogallol ($C_{18}H_7Cl_{11}O_{10}$) and leucogallol—products of chlorination of pyrogallol, under carefully regulated conditions—in combination with the furfural-yielding complex. The combination with the chlorine is attended by molecular hydration, in consequence of which the chlorinated lignone is split off, more or less, from its condensed union with the cellulose.

As in the preceding case, therefore, we are dealing with a reaction which, though perfectly definite and characteristic of constituent groups of the parent molecule, cannot be interpreted in terms of these groups without introducing hypothetical considerations. The reaction will be discussed subsequently from this point of view. In the mean time it is sufficient to point out that the reaction is uniform in its empirical features, that these may be quantitatively studied, giving what we may term the constants of chlorination of the fibre-substances, viz. (1) the chlorine *combining* with the hexene groups of the fibre-substance as quinone chloride; (2) the chlorine combining with hydrogen, and set free as hydrochloric acid. It is found, in the case of jute, that (1) and (2) are approximately equal, and therefore that the reaction is unattended by secondary oxidations of the fibre-constituents to any notable extent.

The following are the details of the methods of estimating these *constants of chlorination*.

(1) *Volume of chlorine disappearing in chlorination.*—The method of observation is fully described in J. Chem. Soc. 1889, 169. The fibre-substance is prepared in the usual way, by previously boiling for 10-15 minutes in dilute alkaline solution

(1 p.ct. NaOH). This treatment is carried out in duplicate, one of the treated specimens being washed off with water, dilute acid (acetic), and finally water, and dried, for the estimation of the loss of weight in the alkaline treatment. (*a*) The second portion, after thorough washing from the alkaline treatment and finally with distilled water, is squeezed so as to retain a minimum of water, introduced into a glass bulb of extremely thin walls, and sealed off. (*b*) The bulb is carefully introduced into a bottle previously filled with chlorine gas, collected over warm water and inverted with a glass plate placed on the mouth, and in such a way that a minimum quantity of water is left in the bottle. In any case a suitable quantity of coarsely pounded glass should be introduced with the bulb, in order to prevent the fibre being unduly wetted, which would retard the absorption. The bottle is closed with an indiarubber cork, well coated with paraffin, through which passes a bent tube. Through this tube the bottle is brought into connection with any suitable gas-measuring apparatus permitting accurate measurement of the vacuum formed in the bottle as the reaction proceeds. All parts of the apparatus being brought to a constant temperature, the bulb containing the fibre is broken by a blow against the sides of the bottle. The chlorine is absorbed with rapidity, and observations of the absorption are made from time to time. If an apparatus such as a Lunge's nitrometer is used, the apparatus is adjusted at its extreme mark, i.e. full of air. As the volume of gas in the reaction flask shrinks, the liquid levels in the measuring apparatus are adjusted in the usual way. The reaction may be considered at an end when no further absorption is noted during an interval of 10 minutes. It is advisable to insert a stopcock between the reaction bottle and the measuring apparatus, so that the latter may be cut off after each observation of volume.

In calculating from the observed numbers, it is only necessary to note that the gas disappearing in combination is a volume observed under the conditions of the experiment; it is corrected therefore for temperature, barometric pressure, and the partial pressure of aqueous vapour, and the corrected number reduced to the *weight* of the chlorine taking part in the reaction. This weight is 16-17 p.ct. of the weight of the fibre-substance.

The quantity of the latter which may be conveniently taken is 1-2 grms.; the quantity of gas, measured under ordinary conditions, required for 2 grms. of the lignocellulose is 100-120 c.c. It is advisable to take a reaction bottle, of capacity equal to twice this volume.

The errors of experiment in such a determination are not very considerable; they may be minimised by keeping the reaction bottle submerged in water of constant temperature, and shielding it from the light, to prevent interaction of the chlorine and water; also by observing the precautions usual in the measurement of gas volumes.

(2) *Determination of HCl formed in the reaction.*—The quinone chloride formed in the reaction is slightly soluble in water, but almost insoluble in a solution of common salt (20 p.ct. NaCl). By washing the chlorinated fibre with a neutralised salt solution, the hydrochloric acid may be removed. The reaction bottle being disconnected, the salt solution is poured down the sides, the fibre and bottle being further washed once or twice with the salt solution. Residues of chlorine are removed by passing a current of air for a minute or two through the solution, which is then treated with standard alkali in the usual way.

The chlorine converted into hydrochloric acid is, in the case of jute, approximately one half the total chlorine entering into reaction, i.e. from 8-8·5 p.ct. The reaction appears, therefore, to be one of simple substitution of hydrogen. In the case of other lignocelluloses, yet to be examined, it is found to be in excess of the half, as a result of oxidising actions. This point should be borne in mind.

(3) *Control observations.*—(a) The chlorine in combination in the chlorinated fibre may be directly estimated by any of the standard methods by which the 'organic' molecule is broken down and the chlorine liberated as hydracid.

The chlorinated fibre itself is, for obvious reasons, somewhat difficult to deal with. The chlorinated product may be dissolved by treatment with pure sodium hydrate, by which treatment it is largely decomposed. To complete the isolation of the chlorine as sodium chloride, the solution and washings are boiled down to dryness, and heated for some time at 200–300° C. An iron dish may be used for this treatment. The soluble chloride is dissolved out and precipitated as silver chloride, in presence of nitric acid.

(b) The cellulose may be isolated in the usual way, by boiling the fibre-substance with sodium sulphite solution, and further treating the cellulose as described, p. 95. The resulting solution and washings of the cellulose may also be employed for the estimation of the chlorine, the organic products being destroyed by oxidation with nitric acid. Sufficient silver nitrate being previously added, the oxidation may be carried out in an open flask attached to an upright condenser.

The chlorination has also been studied in a different way. (1) for the estimation of the total chlorine combining ; and (2) for proving that no destructive oxidation takes place.

Weighed quantities of the fibre-substance were chlorinated (a) after boiling in water, (b) after boiling in 1 p.ct. NaOH solution. Duplicate specimens were weighed after these treatments and without chlorination, and the statistics are worked out upon the weights of the fibre-substance *after* treatment. After chlorination the products were transferred to a bell jar containing an ample supply of solid potassium hydrate ; the vessel was exhausted, and the fibrous products left for some days. After a second similar exposure *in vacuo* over solid KOH, and with addition of sulphuric acid (in separate vessels), the specimens were exposed for a short time to a temperature of 100° in a water-oven and weighed. The combined chlorine was then estimated. The following are the results:

	(a)	(b)
Weight of fibre-substance chlorinated	1·912	1·612
Combined chlorine determined	0·142	0·153
	2·054	1·765
Chlorinated fibre obtained	2·038	1·763
Loss due to oxidation	0·016	0·002

These results, in conjunction with independent observations of the hydrochloric acid formed, further confirm the conclusion as to the simplicity of the reaction.

The percentages of chlorine combining—viz. (a) 7·4, (b) 9·4—vary on either side of what may be taken as the mean number, viz. 8·0 p.ct. In both cases it was no doubt impossible, under the conditions of the after treatment, to entirely remove the HCl. The difference in favour of (a) shows the importance of the preliminary treatment with the alkali; without this the chlorination is incomplete.

Estimations of Secondary Constituents. — (a) *Methoxyl* ($O.CH_3$) is estimated by the now well-known and, in fact, standard method of Zeisel. The fibre-substance is boiled with hydriodic acid; the resulting methyl iodide is washed in an apparatus of special construction, to remove traces of hydriodic acid, and passed into an alcoholic solution of silver nitrate, with which it reacts to form silver iodide. A constant current of carbonic anhydride is passed into the reaction flask and through the entire apparatus, so that the methyl iodide may be continuously carried forward and quantitatively decomposed.

The calculation from silver iodide to methoxyl is, of course, simple ($AgI = O.CH_3$).

This constituent of the lignocellulose molecule we have every reason to regard as a characteristic *constant*, and its determination is therefore of importance. Its constitutional relationship will be discussed subsequently.

(b) *The $CO.CH_2$ residue.*—In a number of decompositions

of the fibre-substance, acetic acid is formed. The maximum yield is obtained in the process of decomposing by digestion with dilute nitric acid. As it is also formed in considerable quantity by dissolving the fibre in sulphuric acid, in the cold, and obtained from the solution by diluting and distilling, it must, in such case, be regarded as a product of hydrolysis, and not of oxidation of the lignocellulose. The problem of its constitutional relationship will be discussed in due course. The estimation of the acid in the latter case need not be dealt with, but it is necessary to describe the method by which it is estimated after being liberated by the nitric acid treatment. The weighed quantity of fibre is placed in a flask and treated with 4 times its weight of 5 p.ct. nitric acid. It is digested for 5-6 hours at 90°, with the flask heated in a water-bath, and attached to an upright condenser. The lignone being entirely resolved, the acid solution is poured off, and the fibrous residue washed with hot water. It is then digested with 5 p.ct. of its weight (original fibre) of sodium carbonate dissolved in a small quantity of water, which completes the removal of the lignone derivatives from the residual cellulose. The solution and washings are added to the original acid liquid. It is now necessary to destroy the residues of nitric acid before distilling for the volatile acid. A small quantity of sulphuric acid is added to the liquid, in a flask, which is then digested for some hours upon metallic iron. The solution is then boiled for some time, with the flask attached to an upright condenser. Urea is then added, and the solution distilled, taking care that the sulphuric acid is present in slight excess. For the complete removal of the acetic acid it is necessary to distil over as much of the contents of the flask as possible, and to repeat the distillation at least twice, adding a certain volume of water to the flask, and taking over an equal volume of the distillate.

The distillate is then made up to a definite volume and

titrated. A portion is drawn off and tested for nitric and nitrous acids. If free from these acids, the titration number may be taken as representing the acetic acid. Otherwise a fraction of the distillate must be further treated for the elimination of the nitrogen acids. For this purpose it is acidified with sulphuric acid, and digested for some hours with a 'copper-zinc couple.' The solution is then poured off, and distilled, as before described, from a slight excess of sulphuric acid.

These determinations of acetic acid are destined to contribute, in an important way, to the solution of constitutional problems, and the student should master the details of the somewhat laborious process above described.

In certain cases, other volatile acids may be formed. It is advisable to control the results by testing the distillate qualitatively, and should there be indications of the presence of other acids, e.g. formic acid, a fraction should be redistilled from chromic acid, and the distillate again titrated.

The distillates also may be divided, a portion being titrated, and the acid in a portion converted into silver salt in which the silver is determined.

Having thus described in general terms the constituent groups of the typical lignocelluloses, and more specially the methods by which they may be quantitatively estimated, directly or indirectly, it is necessary to point out that so far nothing has been said as to the mode of union of these groups in the fibre-substance. The evidence on this side of the subject will be given in due course. It is sufficient, for the present, to remember that we are dealing, on an empirical basis, with the well-ascertained chemical constants of lignification—constant for any given lignocellulose, but varying considerably from member to member of this wide and varied group of plant constituents. It may not be out of place also to insert at this point a caution against the possible inference that the above groups are sharply separated from one another. As the student becomes more familiar with the subject, he will find the constituent groups 'overlapping' in an unmistakable manner, with suggestions of probable genetic connections.

Keeping, however, for the present, to the strictly empirical view, we proceed to the systematic account of the jute fibre as the typical lignocellulose.

The JUTE FIBRE is the isolated bast of plants of the species *Corchorus* (Order Tiliaceæ), an annual of rapid growth, usually attaining a height of 10-12 ft. in the few months required, in the Indian climate, for the maturing of the plant. This great length of stem is attained without branching, and the separation of the bast from the wood and cortex is a manual operation of the simplest kind. The plants, after being cut down, are steeped or retted for a short period in stagnant water; the stems are then handled individually; the wood being broken, the bast is easily stripped and freed by washing from the softened cellular cortex. The fibre is supplied to commerce in long lengths, or strands, representing nearly the full length of the parent stem. As, however, the lower portion, 6-8 ins. from the root upwards, is more or less reticulated, this is usually cut off, and these rejections constitute the jute 'butts' or 'cuttings' largely used as the raw material for special classes of wrapping papers. The textile fibre is of a brown to silver-grey colour in the finer sorts. The individual fibres, or spinning elements (filaments), are complex structures; in cross section they are seen to be bundles of the ultimate fibres, the number of which varies from 7 to 20. The ultimate fibre itself is of short length, 2-3 mm.; it is of circular or polygonal section, with a central canal sometimes nearly obliterated, from the thickening of the cell wall. These bast fibres taper off at their extremities, and are built up by apposition to form the complex filament or bundle.

The fibres or filaments are somewhat matted together in the strands by reason of the great pressure under which the bales are packed, and also in part owing to the presence, in the tissue, of mucilaginous or pectic bodies (parenchymatous residues &c). Jute requires, therefore, a softening treatment as a preliminary to the preparing operations of the spinner. It is opened out from the bales, dusted, and passed through a series of heavy fluted 'breaking' rollers, being simultaneously sprinkled with water and whale-oil. By this treatment the subdivision and drawing of the fibres in the hackling, or combing, and spinning processes is greatly facilitated. For the purposes of laboratory investigation the fibre may be freed from adventitious impurities by boiling in weak solutions of sodium carbonate, washing well to remove soluble matters, and rubbing well in a stream of water, to remove residues of cortical parenchyma.

The bast fibre thus obtained is somewhat harsh to the touch, coloured as described, more or less, and having a certain amount of lustre.

Its specific gravity is 1·436 (Pfuhl), 1·587 after purification by boiling in alkaline solutions (Cross and Bevan).

The following results of proximate analyses of various specimens are given by Hugo Müller, Pflanzenfaser, p. 59.

	Long fibre		Brown cuttings
	Nearly colour-less specimen	Fawn coloured	
Ash	0·68	—	—
Water	9·93	9·64	12·58
Aqueous extract . .	1·03	1·63	3·94
Fat and wax . . .	0·39	0·32	0·45
Cellulose	64·24	63·05	61·74
Incrusting substances and pectic constituents. Difference from 100 .	24·41	25·36	21·29

To compare these results—chiefly for cellulose—with the

authors' results given in the text, the numbers must be calculated to dry substance (i.e. multiplied by 1·1).

The authors have not isolated pectic acid from the fibre-substance proper; but jute 'cuttings' often contain a considerable quantity.

Composition.—The inorganic constituents amount to from 0·8–2·0 p.ct., and are obtained, on burning the fibre, as a brownish-coloured ash, of which the preponderating constituents are silica (35 p.ct.), lime (CaO 15 p.ct.), and phosphoric acid (P_2O_5 11 p.ct.). Manganese is usually present in small quantity (Mn_3O_4 0·75 p.ct.).

The organic portion, or fibre substance proper, varies somewhat in composition, the subjoined numbers representing the mean range of variations:

$$C \quad . \quad . \quad . \quad . \quad 46\cdot0\text{-}47\cdot0 \text{ p.ct.}$$
$$H \quad . \quad . \quad . \quad . \quad 6\cdot3\text{- } 5\cdot8 \text{ ,,}$$

In dealing with the jute fibre *substance* in contradistinction to the jute fibre, the results are referred to the substance taken as dry (100°) and when the result would be seriously influenced, as ash-free.

For 'statistical' purposes, therefore, the fibre-substance may be represented by the empirical formula $C_{12}H_{18}O_9$. There is plenty of evidence for the view that lignification is an intrinsic process of chemical change of cellulose, and it might therefore be inferred that the process is one of dehydration:
$C_{12}H_{20}O_{10} - H_2O = C_{12}H_{18}O_9$.

As an illustration of the superficial meaning of such numerical relationships, we may cite here the results obtained by A. Pears in cultivating the jute plant under the more artificial conditions of growth in a 'hot house' in this country. A normal growth of the plant was secured, in the sense that the seed saved gave a satisfactorily high proportion of germination in the second year of cultivation, and from both cultivations good specimens of the bast fibre were separated in the usual way. The composition of these specimens was determined as follows:

	Fibre grown in 1892	Fibre grown in 1893
C	43·0	43·5
H	—	6·1

In further illustration of the results obtained in these 'artificial' cultivations of the fibre, we reproduce the various numerical

Constituents and reactions	Method	Jute produced in England (1892)	Normal fibre	
Moisture	Drying at 100°	11·4	10·3	—
Inorganic constituents	Ash	1·6	1·2	—
Alkali hydrolysis	1 p.ct. solution NaOH (1) 10 mins. boiling	14·8	8·0	Loss of weight
	(2) 60 mins. boiling	20·0	18·0	,,
'Mercerisation'	20 p.ct. solution NaOH in the cold	12·2	8·0	,,
Nitric acid resolution	5 p.ct. HNO_3.Aq 8 hours at 70°	37·0	37·0	,, Residue oxycellulose
Cellulose	Chlorination &c.	75·2	75·0	—
Chlorine absorption	J. Chem. Soc. 55, 199	13·7	16·6	—
Iodine absorption	Excess of $\frac{1}{60}$ normal solution in KI	6·0	6·0	—
Nitration	Equal volumes of HNO_3 1·5; H_2SO_4 1·82.	130·0	145	—
Ferric ferricyanide reaction	J. Soc. Chem. Ind. 1893	133·0	124	Increase of weight under equal conditions
Thiocarbonate reaction	J. Chem. Soc. 1893, 837	45·0	45·0	P.ct. of fibre undissolved
Carbon percentage	Combustion with CrO_3 and H_2SO_4	43·0	46·5	—

determinations given in the original paper (J. Chem. Soc. 1893, 967).

The preceding table contains the results of a more extended scheme of investigation than is required for special and more practical purposes. The results, however, all have the value of 'constants,' depending as they do upon definite properties of the fibre-substance. It is an amplification of the scheme adopted by Webster (J. Chem. Soc. 43, 23), working in collaboration with the authors; and, again, of that given by the authors in the Reports Col. and Indian Exhibition, 1886.

For the fibre grown in 1893, from the seed saved from the above, the following constants were determined (J. Chem. Soc. 1894, 471):

	C	H
Elementary composition	43·5	6·0
Furfural yield	8·55 p.ct.	
Cl absorption	15·0 ,,	

The chief feature of these results is the preservation of the essential constitutional features of the lignocellulose with such considerable variation from the normal in elementary composition.

Notwithstanding this wide divergence in composition, the fibre-substance showed all the essential characteristics of *constitution* of the ordinary product. The observed difference is, therefore, in the main associated with *hydration*; and lignification is evidently a process which is independent of dehydrating conditions.

The lignocelluloses, however, under normal conditions of growth are progressively dehydrated, and in nearly all cases, therefore, are characterised by high carbon percentage (46–51). These considerations lead up to the general question of the relationships of the jute fibre to water.

Jute Fibre and Water.—Lignocellulose Hydrates.

The *hygroscopic moisture* of ordinary jute varies, under normal atmospheric conditions, from 9–12 p.ct., the variation being, of course, chiefly dependent upon temperature and 'dew point,' or rather the percentage saturation of the air with aqueous vapour.

In an atmosphere saturated at ordinary temperatures jute takes up 23 p.ct. of moisture.

The hydration of the fibre-substance, in the more permanent sense of definite combination with H_2O molecules, is determined under conditions which will appear in the succeeding sections of the subject.

Solutions of Lignocellulose.—The jute fibre is attacked and dissolved by the solvents already described under 'Cellulose' (p. 8), viz. :

(1) Zinc chloride—concentrated aqueous solution.

(2) Zinc chloride—solution in HCl ; and

(3) Cuprammoniun solutions.

From these solutions the lignocellulose is precipitated, on dilution (1 and 2) or acidification, as a gelatinous hydrate ; the precipitation is, however, incomplete—the proportion remaining in solution varying from 15–25 p.ct., according to the conditions of solution. There is, however, no difference in reactions between the soluble and insoluble fractions, and on ultimate analysis both are found to have the empirical composition of the original fibre-substance. Although, therefore, the lignocellulose is a complex of various groupings, it behaves in this respect as a homogeneous product, and the bond uniting the groups together is not resolved by simple hydrolytic agencies (see *infra*, p. 134).

In the case of the $ZnCl_2.HCl$ reagent the fibre-substance is progressively hydrolysed on standing. This is illustrated by the following determinations of the proportion of the lignocellulose reprecipitated from such solution.

(*a*) Precipitated at once : ppt. 78·4 p.ct. of the original.

(*b*) After standing 16 hours : ppt. 29·4 p.ct. of the original.

Qualitative Reactions and Identification of the Lignocelluloses.—Whereas the reactions of the celluloses are mostly negative, jute (and the lignocelluloses generally) is distinguished by a number of characteristic reactions. In addition to those already described as admitting of quantitative estimations, the following are the more important :

(1) *Salts of aniline* (and many of the aromatic bases), in aqueous solution, colour the fibre a deep golden yellow.

(2) The *Coal-tar dyes* generally combine freely with the lignocelluloses. In 'staining' sections of plants and parts of plants for microscopic observation, the lignocelluloses are dyed by the majority of soluble coal-tar dyes. Their 'affinities' for colouring matters are, in fact, similar to those of the animal fibres, silk and wool, although differing radically from them, not only in constitution, but in containing no nitrogen (NH_2 groups).

(3) *Phloroglucinol*, in hydrochloric acid, gives the deep magenta colouration characteristic of the pentaglucoses. The reagent is prepared by dissolving the phenol to saturation in HClAq (1·06 sp.gr.).

(4) *Iodine* is absorbed from its solutions in potassium iodide in large quantity, colouring the fibre a deep brown.

(5) *Chlorine* combines with the fibre with avidity, as already described ; the chlorination is made evident by treatment with sodium sulphite solution, which develops a deep magenta colouration. This reaction is *characteristic*. Conversely, the fibre-substance may be employed as a reagent for the identification of chlorine, or may, in certain cases, be used for absorbing the gas.

(6) *Ferric chloride* colours the fibre-substance to a dark greenish tint—the reaction being due to traces of tannins.

(7) *Ferric ferricyanide*—the red solution obtained by mixing together ferric chloride and potassium ferricyanide in equivalent proportions—gives a highly characteristic reaction (subsequently described in detail), the fibre-substance rapidly decomposing the compound to 'Prussian blue,' the pigment being taken up

by the fibre-substance in very large quantity (50 p.ct. of its weight).

(8) *Chromic acid*, in aqueous solution, combines with the lignocellulose, and is then very slowly reduced to the intermediate oxide ($CrO_3.Cr_2O_3$).

(9) *Potassium permanganate* is rapidly reduced, the MnO_2 produced colouring the fibre a deep brown. After treatment with sulphurous acid, which removes the oxide, the lignocellulose will be found to have been bleached by the treatment. On repeating this treatment once or twice, with dilute solution of $KMnO_4$, the lignocellulose is obtained of a cream or greyish-white colour, the loss of weight sustained in the bleaching being small (2-4 p.ct.).

Compounds of Jute Lignocellulose.—The fibre-substance itself being a complex or compound cellulose, and susceptible of decomposition (*a*) by hydrolytic treatment—in which, however, the union of the constituent groups is preserved—and (*b*) by reagents which selectively attack the constituent groups, it is obvious that we are limited in the preparation of compounds which may be regarded as compounds of the lignocellulose molecule as a whole. We shall first describe those which result from reactions of the OH groups of the lignocellulose. These are more active than in the celluloses. We have already pointed out that the fibre combines freely with colouring matters. The phenomena of dyeing being now well established, as the result of interaction of salt-forming groups in fibre-substance and colouring matter, i.e. a species of 'double salt' formation, we may deduce from the considerable and very general 'affinity' of the lignocelluloses for the coal-tar colouring matters, that they contain OH groups of both acid and basic function, and much more disposed to reaction than those of the celluloses.

Absorption of acids and alkalis from dilute aqueous solutions.—

This phenomenon, already described as a property of the celluloses, is more pronounced with the jute fibre. The following absorptions have been determined by the authors from normal solutions of the respective reagents:

Normal hydrochloric acid.—(a) Fibre digested with 8 times its weight of solution, 10 minutes at 15° C.; (b) with 20 times its weight.

	(a)	(b)
HCl absorbed	0·85	1·1 p.ct. on fibre-substance

Normal sodium hydrate.—Fibre digested with 20 times its weight of solution.

	(a)	(b)
Na_2O absorbed	3·0	3·6 p.ct. on fibre-substance

It is to be noted that the molecular ratio of the absorptions is approximately that observed in the case of cotton, viz. 3HCl : 10 NaOH.

The hydrolysing action of the alkalis (a) and non-oxidising acids (b) may be regarded as an extension of this phenomenon. (*a*) The alkalis and alkaline compounds in aqueous solution attack the fibre-substance in the ratio of their hydrolysing and saponifying activity; and, as in the action of the solvents previously described, the lignocellulose is attacked as a whole. In the systematic comparison of the vegetable fibres (i.e. compound celluloses) it is important to determine their relative resistance to alkaline treatments under standard conditions. It is usual, for this purpose, to determine the loss of weight sustained by the fibre on boiling with a 1 p.ct. solution of sodium hydrate (i) 5 minutes, (ii) 60 minutes. Under this treatment jute loses on the average—

(i)	(ii)
8·0	15·0

No change in the composition of the fibre-substance is occasioned by the treatment, the portion dissolved showing the essential characteristics of the original fibre. It is precipitated in part on acidifying the solution, and the gelatinous precipitate gives the characteristic reactions of the original fibre. The fibre is further but slightly affected in structure and physical properties by the treatment.

At temperatures considerably above the boiling point the action of dilute solutions of the alkaline hydroxide (1-3 p.ct. Na_2O) takes a different course ; the non-cellulose is attacked and converted into soluble derivatives, and the fibre-bundles are more or less disintegrated. Such processes are, in fact, used on the large scale for the preparation of paper-making pulp (cellulose) from the lignocelluloses.

(*b*) On digesting the fibre with dilute solutions of the mineral acids at 60-80°, the lignocellulose is again progressively dissolved, the loss of weight sustained by the fibre being proportional to the hydrolysing activity of the acid, and to the conditions of the digestion. In this case also the lignocellulose is attacked as a whole, the insoluble fibrous residue preserving the essential characteristics of the original fibre. The dissolved portion may be isolated—when sulphuric acid (5-7 p.ct. H_2SO_4) is used as the hydrolysing acid—by neutralising the solution with barium carbonate, filtering, and evaporating to dryness. The soluble modification of the fibre-substance is obtained as an amorphous, brown, gummy solid, having the same empirical composition as the original fibre.

On prolonged digestion with the dilute acids (5-7 p.ct. H_2SO_4) the loss of weight sustained by the fibre approximates to a limit at about 30 p.ct. As a result of the treatment, the fibre is disintegrated, the residue being obtained as a mass of brittle fragments. It is to be noted that the disintegration is not a progressive dissection of the ultimate fibres—such as results from the alkaline digestions above described—but is the result of a change in the physical properties of the fibre-substance itself; **the** disintegration which ensues **is** characterised by fracture of the fibre-bundles or filaments.

The following results of particular experiments may be cited :

Fibre digested with 7 p.ct. H_2SO_4.
   (1) 18 hrs. at 60–80° : loss of weight, 12·0 p.ct.
   (2) 12 hrs. at 80–90° :   ,,   ,,   9·7 ,,
   (3) 42 hrs. at 80–90° :   ,,   ,,   23·0 ,,

The investigation of the products from (3) gave the following results :

(*a*) *Soluble.*—Isolated by neutralising with $BaCO_3$. Evaporation, solution of residue in alcohol, evaporating solvent and drying at 105° gave, on combustion :

			Calc. $C_{12}H_{14}O_7$
C	46·29	46·08	47·05
H	5·75	5·95	5·88

This substance gave with Cl the characteristic quinone chloride, and on boiling with hydrochloric acid, furfural.

On adding phenylhydrazine acetate to the concentrated solution of the product and heating at 90–100°, an osazone is formed: it separates as a coagulum of characteristic greenish-yellow colour. After washing and drying, the product may be purified by solution in toluene; from which solution satisfactory crystallisations are obtained. A series of these compounds has been obtained with melting points ranging from 110–130°, and containing from 9–10 p.ct. nitrogen. Their relationship to the original fibre-substance has not yet been determined.

(*b*) *Insoluble.*—The brittle fibrous residue gave with chlorine the characteristic reaction, and the cellulose isolated in the usual way amounted to 75 p.ct. of the weight of the product.

The action of the acids proceeds, as stated, **to a** limit which is determined by the concurrent effects of condensation. If the fibre be then washed and boiled for a short time in alkaline solution, it is again rendered susceptible of attack by the hydrolysing acids, with further conversion into soluble derivatives.

If the acid solutions are boiled, the dissolved product is

decomposed with formation of furfural and acetic acid; when formed at temperatures below 70°, it may be regarded as a soluble hydrate of the lignocellulose, or more correctly a derivative of low molecular weight in which the characteristic groupings of the parent molecule are preserved.

Concentrated solutions of the alkaline hydrates.—The structural changes produced in the fibre by treatment with solutions of caustic soda of 'mercerising' strength (15-30 p.ct. NaOH) are remarkable. The fibre-bundles are resolved more or less; the cell wall of the individual fibres undergoes considerable thickening, such that the central canal is almost obliterated. The visible effects of these changes of minute structure are (1) a shrinkage in length of the strands of fibre (15-20 p.ct.); (2) a considerable refinement of the spinning units or filaments; (3) the filaments have a wavy or crinkled outline, resembling that of wool.

The following experimental results may be cited: 80·35 grms. fibre (air-dry) treated with 300 c.c. of 25 p.ct. solution NaOH, six hours in the cold; washed, acidified, washed and dried; weighed, air-dry, 75·5 grms. Loss of weight, 6 p.ct. Shrinkage in length, from 4 ft. to 3 ft. 8 in., i.e. 17 p.ct.

An extended series of experiments upon normal specimens of the raw fibre, with varied conditions—e.g. concentration of alkali, 15-33 p.ct. NaOH; duration of treatment, 5 minutes to 48 hours—showed an average loss of weight of 7·5 p.ct., with slight variations only on either side of this mean number. The same specimen of jute lost 11·9 p.ct. in weight on digesting 48 hours in more dilute alkali (6·5 p.ct. NaOH), and 11 p.ct. on boiling for 5 minutes in alkali of the same concentration.

The cellulose constants of the fibre are unaffected by the treatment.

The chemical changes are more complex than with the celluloses, for reasons which will appear when the constitutional relationships of the constituent groups of the fibre-substance are discussed. Empirically the results of the treatment are as follows :

A certain proportion of the lignocellulose is dissolved ; but the dissolved portion, as well as the fibrous residue, gives the characteristic reactions of the original. When the latter is chlorinated, and the cellulose isolated in the usual way, the percentage yield is found to be unaffected by the treatment. The character of the cellulose is somewhat altered, however, as it is obtained in continuous strands ; and when dried, the filaments of jute cellulose have a certain amount of coherence. So far there is a general resemblance to the changes produced in cotton cellulose on mercerising, i.e. the effects are chiefly *hydration changes*. The differences, on the other hand, are brought into evidence when the alkali-lignocellulose is exposed to the action of carbon disulphide.

The *thiocarbonate reaction* which ensues is of a remarkable character. The lignocellulose is gelatinised more or less in the reaction, but on treatment with water it is not dissolved to a homogeneous solution, but swells up enormously, the hydration proceeding to almost indefinite limits. The following results of an experiment may be cited :

4·5 grms. fibre, treated with excess of 12 p.ct. solution of NaOH, squeezed, and exposed to CS_2 (2·0 grms.) in a stoppered bottle 24 hours. On treatment with water, the gelatinised fibre occupied a volume of 300 c.c. ; and for separation of the undissolved fibre, dilution to 750 c.c. was necessary. The following determinations were made :

  (*a*) Undissolved fibre . . . . . . 42·7
  (*b*) Dissolved—reprecipitated by HCl . . . 43·3
  (*b'*) Soluble after acidification . . . . . 14·0
  (*a*) Gave the reactions of the original fibre-substance.
  (*b*) Gave only a slight reaction with Cl and Na_2SO_3.
  (*b'*) Consisted mainly of the furfural-yielding constituent.

The following results of particular experiments are of interest :

(1) Ten grms. raw jute (with 10 p.ct. moisture), purified by boiling in dilute solution Na_2CO_3; washed, squeezed, and placed in bottle with 4 grms. CS_2. When evenly diffused, treated with 25 c.c. of 15 p.ct. NaOH and left 48 hours.

Insoluble product (after purifying), 7·053 grms. (dry); yield on 9 grms. (dry), 78·4 p.ct. Gave 4·1 p.ct. furfural on distillation (HCl).

(2) Conditions exactly as in (1), with which it was comparative in regard to effect of varying the reaction; viz. in this case the jute was *first* treated with the NaOH, and afterwards sealed up with 4 grms. CS_2.

Insoluble product, 7·004 grms., 77·8 p.ct. The variation in question was therefore without effect. The *filtrate* from the insoluble residue was treated with zinc acetate in excess, which has been found to precipitate the dissolved fibre-constituents. The precipitate was then decomposed with HCl in excess, and the now insoluble fibre-products washed, purified, dried, and weighed. Weight, 0·980 grm. On distillation with HCl this gave 0·031 grm. furfural; the filtrate from this insoluble product gave none.

The undissolved fibre was chlorinated, and the cellulose separated in the usual way: 5·728 grms. obtained, i.e. 81·8 p.ct. on product, or 63 p.ct. of original lignocellulose. The results are as follows:

The furfural-yielding groups have been attacked; the total yield of the aldehyde is reduced by 50 p.ct., the reduction falling chiefly upon the portion *hydrolysed* and *dissolved*. The a-cellulose is unaffected, the keto R. hexene groups also. The portion dissolved appears to be the β-cellulose and the furfural-yielding constituent of the lignone.

(3) The above conditions were maintained, varying the duration of action of the alkali; it was left 48 hours before adding the CS_2. The results were:

Insoluble fibre, 6·82 grms. = 75·8 p.ct.
Giving cellulose (after Cl &c.), 5·76 grms. = 64 p.ct. of the *original*.

Under constant conditions as regards the reagents, the results are therefore independent of the mode of carrying out the reaction.

The reaction requires further investigation, as it appears capable of throwing light upon the actual mode of union of the constituent groups in the lignocellulose.

The results obtained are, however, by no means constant, but vary considerably with variations in the conditions of treatment. The causes of these variations lie in the complex character of the lignocellulose. In the celluloses alcoholic characteristics predominate; in the lignocelluloses the presence of phenolic OH groups, of $CH_2.CO$ and $CH_2.CO.O$ groups, and of ketonic oxygen (see p. 137) must largely modify the alcoholic functions of the cellulosic OH groups. They are, in fact, in condensed union with these more negative groups, and this union is only partially resolved in certain directions, and further cemented in others, by the alkaline treatment. It is probable that the alkali may have the effect of further condensing or synthesising the aldehydic groups.

It is evident, on the other hand, that the entire molecule is opened up for the entrance of water molecules, and the prominent result of the reaction is the consequent hydration of the lignocellulose. It is this aspect which leads us to describe the reaction under the heading of the 'Compounds of the Lignocellulose.' The attendant hydrolysis is a secondary result which will be referred to subsequently.

Compounds of the Lignocellulose with Metallic Salts.—We are still dealing with those synthetical reactions of the lignocellulose (OH groups) which take place in presence of water. The resulting compounds are necessarily of a variable and ill-defined character, owing to the complex nature of the lignocellulose, the tendency to selective reaction with its constituent groups, and to consequent partial hydrolysis.

The combinations already described as taking place with the alkalis and non-oxidising acids are seen to be of a feeble and transitory character. With many of the salts of the heavy metals the reactions are more pronounced, but they are also of too indefinite a character to require more than a passing notice. Generally we may regard reaction as occurring only with such as undergo pronounced dissociation on solution in water, which dissociation is exaggerated by the fibre-substance. The lignocelluloses present merely a particular case of the general theory of the action of 'mordants,' its combinations with the 'mordanting' oxides being similar to those of the cellulose, differing only in the higher proportions of the oxides taken up.

There is one reaction, however, of a specific character, already alluded to, which merits description in detail—that is, the interaction of the *lignocellulose and ferric ferricyanide*.

This reaction has been described in a paper by the authors in the J. Soc. Chem. Ind. 1893, the experimental portion of which is now reproduced, with certain alterations of minor import.

A REACTION OF THE LIGNOCELLULOSES AND THE THEORY OF DYEING.—The red solution obtained by mixing aqueous solutions of ferric chloride and an alkaline ferricyanide, which may be regarded as containing ferric ferricyanide, reacts as is well known with the more easily oxidisable 'organic' compounds, oxidising them and being itself reduced to the lower mixed cyanides, i.e. Prussian blue and similar compounds (Watts' Dict. ii. 248). The reactions of this solution with the lignocelluloses, and notably jute, are remarkable and characteristic. Not only is the conversion into the blue pigment very considerable in proportion to the weight of fibre substance, but the colouring matter is deposited within the fibre in such a way as to give the effect of a homogeneous dye. Thus if this particular fibre, suitably purified by previously boiling in dilute alkaline solutions and washing, be plunged into a $\frac{1}{2}$ decinormal solution of the reagent (prepared by mixing decinormal solutions of the reagents in equal volumes) the fibre-substance rapidly dyes to an intense blue-black with a gain of weight of from 20-50 p.ct.; and the dyed fibre examined under the microscope is of an intense transparent blue with all the characteristics, that is to say, of a 'solid solution' of the colouring matter.

The following is a brief account of the results of quantitative observations. In our first series of observations we used a solution of ferric chloride containing Fe_2Cl_6, equivalent to 0·0976 Fe_2O_3 in 10 c.c., and an equivalent solution of ferricyanide. Five equal portions, each weighing 2·762 grms., of jute fibre were treated with the mixture of the above solutions in equal volumes, in the proportions and with the results given in the subjoined table.

Vols. of solution		Increase of weight	Increase per cent. of fibre	Fe_2O_3 added as Fe_2CN_6	Fe_2O_3 fixed as blue cyanide
Ferric chloride	Ferricyanide				
C.c.	C.c.				
10	10	0·135	6·7	0·195	0·123
20	20	0·489	17·7	0·390	0·230
30	30	0·768	27·8	0·585	0·373
40	40	1·025	37·1	0·781	0·510
50	50	1·161	42·0	0·976	0·601

It will be noticed that the Fe_2O_3 fixed by the fibre is proportional to the quantity taken, viz. approximately two-thirds in each experiment; but the corresponding increase in weight of the fibre due to the blue cyanide fixed is somewhat variable. The cyanide, in fact, is shown by analysis to vary in composition slightly in the ratio Fe : CN, more considerably no doubt in its condition of hydration. The ash left on ignition of the dyed fibre we find to contain no soluble basic constituents, therefore no K_2O appears to be fixed.

Analysis of blue cyanide fixed by the fibre.—A specimen of fibre, dyed blue under the above conditions and weighted to approximately 20 p.ct., was analysed, the total Fe being determined as Fe_2O_3, the CN as NH_3.

$Fe_2O_3 = 6\cdot1$ p.ct. $= 4\cdot27$ Fe ; $N = 3\cdot15$ p.ct. $= 5\cdot85$ CN.

Whence the ratio—

$$\frac{4\cdot27}{65} : \frac{5\cdot85}{26} = 0\cdot076 : 0\cdot225$$

or—

$$Fe : CN = 1 : 3.$$

A portion of the fibre was further treated until the increase of weight amounted to 40 p.ct., and then analysed with the following result :

$Fe_2O_3 = 14\cdot0$ p.ct. $= 9\cdot8$ Fe $N = 9\cdot3$ p.ct. CN.

$$Fe : CN = \frac{9\cdot8}{56} : \frac{9\cdot3}{26} = 0\cdot175 : 0\cdot357 = 1 : 2.$$

By the continued interaction of the fibre-substance and the ferric ferricyanide the Fe''' appears to be deoxidised, and in exhausting the fibre-substance with dilute alkali in the cold, ferrocyanide is dissolved, as of course is to be expected.

It is obvious that we are dealing with an aggregate and the product of a mixed reaction ; the ratios $Fe_2(CN)_6$ and $Fe_3(CN)_6$ determined as above for the product in successive stages are therefore not to be taken as more than approximate indications of the composition of the colouring matter deposited in the fibre.

The mechanism of the reaction and the question of the composition of the resulting blue cyanide are further elucidated by the following experiments :

Equivalent solutions were prepared as above, and in two experiments with equal weights of fibre the solutions were mixed in the proportions :

A. 3 of ferric chloride to 2 of ferricyanide ;
B. 2   ,,   ,,   3   ,,

The results observed, which were somewhat unexpected, are given below.

Increase of weight in fibre	Analysis of dyed product		
	Fe_2O_3	N	Ratio Fe : CN
P.ct.	P.ct.	P.ct.	
20	9·6	4·8	$Fe_2(CN)_6$
17	8·4	4·0	$Fe_4(CN)_{11}$

The N in these analyses was determined by the Kjeldahl method, in the previous cases by the soda-lime combustion. It appears, therefore, that the blue cyanide deposited in the earlier stages of the reaction is fairly constant in composition notwithstanding considerable variations in the proportions of the reagents in the solution from which it is abstracted by the fibre ; and the ratio of Fe to CN in the fibre cyanide complex is approximately 1 : 3. By the last-cited experiment it is also shown that the rearrangement of the Fe and (CN) which takes place within the fibre-substance is to a certain extent independent of the condition of the Fe in the solution, i.e. whether added as basic Fe or as in the ferricyanogen complex.

The main points of the reaction have been sufficiently elucidated by the results described, and it is unnecessary to put on record a large number of quantitative results which merely confirm those already given. It should be noted, however, that the limit of the reaction has not been investigated : on occasions we obtained increases of weight of 50 and even 80 p.ct., but at this degree of loading the natural lustre of the fibre had given way to the dull and dusty look of a fibre weighted to excess.

We have now to consider the mechanism of the reaction which has been in some measure elucidated by further experiments.

To explain it as the result of reduction of the ferric iron

either of the chloride or the ferricyanide by the fibre-substance will be found to be inadequate. The following experiments show that either reagent taken singly is but slightly affected by prolonged contact with the fibre-substance.

Equal weights (2·835 grms.) of the purified fibre were placed—

(a) In a solution (30 c.c.) of $FeCl_3$ — 1·6 grm. per 100 c.c.
(b) In a solution (30 c.c.) of K_3FeCy_6 — 3·3 grms. per 100 c.c.
(c) In a solution prepared by mixing the above (30 c.c. of each).

After standing some hours (a) and (b) were squeezed and interchanged, and left some minutes. The fibre from each was then washed off, dried, and weighed with the following results:

—	Weight of dyed fibre	Increase of weight	Colour
		P.ct.	
(a)	2·931	3·3	Indigo-blue
(b)	2·846	0·3	Medium-blue
(c)	3·550	25·2	Blue-black

It appears, therefore, that the lignocelluloses absorb but little[1] oxide from a neutral solution of ferric chloride, and there is only partial reduction of the oxide so fixed: and also that ferricyanide is slightly reduced by the fibre-substance in neutral solution and without sensible combination with the ferricyanogen or ferrocyanogen group.

The reaction in question is therefore specific as between the ferric ferricyanide and the fibre-substance.

That the formation and fixation of the blue product is not the result of reduction in the liquid is further shown by the fact

---

[1] The maximum we have observed to be taken up from a normal solution of the chloride is 0·4 p.ct., and that after 48 hours' immersion.

that it is not appreciably affected by the presence of oxidising agents such as chromic acid.

The alternative conclusion is that it is due to a coagulation or precipitation of the ferric ferricyanide by the fibre-substance, in the first instance, followed by a rearrangement of its constituents by specific combination with the fibre constituents.

Collateral evidence in support of this view is afforded by the behaviour of the ferricyanide with another typical colloid, viz. gelatin. Solutions of gelatin give with the ferric ferricyanide a voluminous coagulum of a greenish colour, and the reaction is approximately a quantitative one, but of course depending to some extent upon the conditions of precipitation. The following relations were determined:

One series of experiments—

A white gelatin (containing 16·5 p.ct. hygroscopic moisture and 2·86 p.ct. ash constituents) was weighed out in quantities of 2 grms. (=1·613 dry and ash-free gelatin) and dissolved. The solutions were variously diluted and treated with half decinormal solution of the ferric ferricyanide added from a burette.

(1) To completely precipitate in the cold 23 c.c. were required.

(2) To completely precipitate at 50° C. 24·5 were required

(3) To a third quantity 32 c.c. of the ferricyanide solution were added.

The precipitates were collected, dried, and weighed; the weights were: (1) 1·990; (2) 2·031; and (3) 2·077 respectively. The mean weight 2·032 shows an increase of weight of 21·6 p.ct. An estimation of Fe_2O_3 in the product gave 6·0 p.ct., the proportion being rather lower than in the case of the fibre-cyanide product, which, with the same gain in weight, contains 7 p.ct. Fe_2O_3. The coagulum is in this case, however, only slightly blue, darkening gradually on standing. On treatment with a reducing agent such as dilute sulphurous acid, the coagulum swells to a deep blue transparent jelly.

This interaction of gelatin and ferric ferricyanide is therefore rather of the character of a simple coagulation or combination of colloids by dehydration.

We have reason for assuming a similar relationship between the fibre-colloid and the ferricyanide as the first cause of the precipitation. The conversion of the colourless into the coloured cyanide may be then accounted for by what we know of the constitution of the fibre-substance.

We have in this complex all the conditions (1) for a deoxidation of Fe''', (2) for union with ferric and ferrous oxides, and (3) combination with HCN.

Such changes as would be determined by these relations, when brought into play, are of the minor order and consistent with the characteristics of the product, i.e. an intimate molecular union of the complex fibre-substance, slightly oxidised at the expense of ferric oxide, and the ferroso-ferric cyanide.

Further investigation has confirmed the interpretation given in the communication which is reproduced above. It is evidently a reaction in which the entire fibre-substance takes part. The ferroso-ferric cyanide being a saline compound, and the lignocellulose containing both acid and basic groupings in combination, and being in that sense analogous to the inorganic salts, the reaction may be regarded as in the main a species of double-salt formation. In this respect it stands on the same footing as the majority of dyeing reactions. But in the formation of the blue ferroso-ferric cyanide from the red ferricyanide the special chemistry of the lignocellulose comes into play. It may be fairly assumed that the deoxidation of the ferric oxide is due to aldehydic groups, the fixation of hydrocyanic acid may be referred to aldehydic and ketonic oxygen, the fixation of ferroso-ferric oxide more particularly to the quinone or keto R. hexene groups of the non-cellulose; the resulting combinations being, however, rather of a 'molecular' character, they reunite to form the coloured lake or 'double salt' represented by the dyed fibre. Without reference, however, to explanations of the mechanism of the reactions, which are for the present more or less hypothetical, the following are the facts to be emphasised in conclusion:

(1) It is a reaction in which the lignocellulose manifests itself as a homogeneous compound.

(2) It is unique in the range of dyeing phenomena both in regard to the formation of the colouring matter by definite chemical reaction, and the very large proportion in which it is fixed by the fibre-substance.

This aspect of the reaction will be found discussed in the original paper (*loc. cit.*), and in a second communication on the subject (J. Soc. Chem. Ind., April 1894) in reply to criticisms by C. O. Weber (*ibid.*, March 1894).

Compounds of Lignocellulose with Negative Radicals.—(*a*) *Lignocellulose esters.*—(1) *Benzoates.*—The interaction of the lignocelluloses with the alkaline hydrates and benzoyl chloride has been only superficially investigated. Fixation of the benzoyl radical certainly takes place; the fibre-substance gains considerably in weight (36 p.ct.), and analyses of the products give results corresponding with the empirical formula $C_{19}H_{22}O_{10}$, which represents the fixation of 1 benzoyl residue upon the empirical molecule $C_{12}H_{18}O_9$ of lignocellulose. This proportion is about one half that of cellulose ($C_{12}H_{20}O_{10}$) under the same conditions. The result accords with the observation of the partial yielding only of the fibre-substance under the thiocarbonate reaction. From both we may conclude that the ratio of alcoholic OH to the total oxygen of the lignocellulose is low in comparison with cellulose.

(2) *Acetates.*—The fibre-substance reacts directly with acetic anhydride at its boiling point; the product shows a considerable gain in weight upon the original lignocellulose, and its reactions are altogether different. It will be obvious that the product in this case will not stand in simple relationship to the parent molecule. In the first place the $CO.CH_2$ groupings of the lignocellulose are liable to further condensation and rearrangement; under the condensing action of the anhydride furfural groupings may be completed, which would then react with the anhydride to form furfuracrylic acid (Berl. Ber. 1894, 286), which, again, would condense with the OH groups of the cellulose; and lastly, the keto R. hexene rings are open to reaction with the anhydride in various ways. In view of these various directions of probable reaction, and the further complications in regard to the analysis of the product, resulting from the presence of $CO.CH_2$ and $CH_2.CO.O$ residues in the lignocellulose molecule, the investigation of the reaction is deferred until the constitution of the fibre-substance itself is elucidated. It cannot then fail to afford confirmatory evidence of great value, both as to the constitution of the constituent groups and their mutual relationships within the molecule.

(3) *Nitrates.*—The 'nitration' of the jute fibre has been studied by O. Mühlhäuser (Dingl. J. 283, 88) and the authors. On plunging the fibre into the well-cooled 'nitrating' acid ($H_2SO_4 + HNO_3$) it is instantly coloured to a dark red. After remaining in the acid for about 5 minutes, evolution of gaseous products is observed. If the fibre be then removed and rapidly washed, the red colour of the nitrate gives place to a golden yellow. The product when dry is found to be somewhat weakened (disintegrated) by the treatment, and harsher to the touch than the original jute. It is, of course, explosive, and takes fire at 160–170°. It is soluble in acetic acid and acetone, and is gelatinised by nitrobenzene and acetic acid.

The following results of experiments (Mühlhäuser) may be cited. The fibre-substance was purified previously to nitration by boiling in alkaline solution (1 p.ct. NaOH), thorough washing, and drying at 100°. The acids used in nitration were of maximum strength.

The results may be given in the form of a table as under :

Ratio of acids in nitrating mixture (by weight)	Proportion of mixture to fibre (by weight)	Duration of nitration (hours)	Yield of nitrate p.ct.	Containing N p.ct. (Eder's method)	
				(1)	(2)
1 H_2SO_4 : 1 HNO_3	10 : 1	1	130	12·10	11·80
2 ,, : ,,	15 . 1	2½	132	12·26	12·04
3 ,, : ,,	15 : 1	3	136	12·03	11·80
2 ,, : ,,	15 : 1	3-4	145	12·03	11·96

The products were purified by exhaustively washing, digestion in dilute solution of Na_2CO_3, and again washing.

An extended series of observations by the authors established the following points :

(1) A gradual increase in yield with increase in duration of exposure to the nitrating acid (at ordinary temperatures) up to 5-6 minutes, the maximum of 145 p.ct. being then attained.

(2) After that oxidation supervenes, soluble products are formed, and the yield of insoluble nitrate diminishes.

(3) The increase in yield again observed on prolonged exposure is a secondary result of the decomposition, alcoholic OH groups being liberated from combination and then taking part in the reaction.

A microscopic examination of the nitrated fibre (Mühlhäuser) showed that nitration by prolonged exposure was attended by resolution of the fibre-bundles into ultimate fibres, and these showed a shrinkage in volume (diameter).

With the view of testing the homogeneity of the product, specimens were exposed to the graduated action of alkaline solutions. One of these, digested 52 hours with sodium hydrate solution (1 p.ct. NaOH), sustained a loss of 22 p.ct. in weight. The insoluble residue, on analysis, was found to contain 12·3 p.ct. N. A second specimen, heated with a 0·5 p.ct. solution of sodium carbonate 3 hours at 90–100°, lost 25 p.ct. of its weight. The residue gave, on analysis, 12·25 p.ct. N.

From these observations, which the authors can fully confirm, the important conclusion was drawn by Mühlhäuser, and may be given in his own words:

'Auch in diesem Falle hatte eine gradweise Abspaltung nicht stattgefunden: die Zerstörung erstreckte sich, wie in allen Fällen, auf das ganze Molecul.' The lignocellulose behaves under nitration as a homogeneous body. It is important to note at this point the convergence to this same conclusion, of the evidence drawn from three independent lines of investigation: (1) the general physiology of the elaboration of the fibre; (2) the resistance of the fibre-substance, so far as regards the union of the constituent groups, to the action of hydrolytic agents; and (3) the homogeneous nature of the products of synthesis, such as the nitrates just described. This evidence compels the view that the fibre-substance is not merely a mixture of cellulose with 'non-cellulose' constituents, but that these are compacted together into a homogeneous, though complex molecule, by bonds of union of a strictly 'atomic' character.

(*b*) **Compounds of Lignocellulose with the Halogens.** (1) *Chlorine.*—The reaction of the fibre-substance with chlorine has been already described. We have now to deal more particularly with the product.

The derivative in question is dissolved in large proportion by treating the chlorinated fibre with alcohol after first washing (to remove HCl) and squeezing. The alcoholic solution may be concentrated by evaporation, and on then pouring into water, the product is precipitated in yellow flocks. On washing, and drying at 100°, and analysing, it gives the following results :

$$\begin{array}{lll} & & \text{Calc. } C_{19}H_{18}Cl_4O_9 \\ \text{C} & 42\cdot82 & 42\cdot85 \\ \text{H} & 3\cdot40 & 3\cdot38 \\ \text{Cl} & 26\cdot83 & 26\cdot69 \end{array}$$

The product, obtained as above described from various specimens of fibre, has been repeatedly analysed. It has been analysed after fractional precipitation from solution in glacial acetic acid; also, after, then again exposing for a long period to an atmosphere of chlorine gas, followed by suitable purification, and always with results in close concordance with the above. It is evident, therefore, that we are dealing with a complex of a definite character. This complex is the *lignin* of earlier observers, but which, in recognition of its ketonic characteristics, is better termed *lignone*. From the fibre-substance exposed to the action of dilute sulphuric acid (5 p.ct. H_2SO_4) for some hours at 60-80° previously to chlorination, a derivative is obtained, having identical characteristics and composition. This further confirms the definite character of the lignone complex, and the resistance of its constituent groups to hydrolytic actions.

The chlorination of the lignocellulose evidently resolves in great measure the union of the lignone to the cellulose residue, as the lignone chloride is largely dissolved away by exhaustive treatment with simple solvents. The residue of chloride which ultimately resists the solvent action gives the characteristic reaction with sodium sulphite, and is probably therefore the same product. The splitting off of the chloride is evidently a secondary result, no doubt an effect of hydrolysis. It is to be noted that the presence of water is essential to the reaction ;

the fibre-substance in the dry state does not react with chlorine even when heated with the gas (60–80°).

The residue, after removing the lignone chloride, is a cellulose, containing, i.e. no 'unsaturated' groups, but yielding, on distillation with HCl, from 4 to 8 p.ct. furfural. It is to be regarded, therefore, as a mixture of a normal cellulose (a) and a cellulose (β) which is readily condensed to furfural. Since the total weight of furfural obtainable from the lignocellulose is not affected by the chlorination, it may be concluded that the 'furfuroids' of the original lignocellulose are in the main associated with the cellulose complex, and from the yields of cellulose, that the principal constituent is this β cellulose constituting 20 p.ct. of the complex. It appears from later researches of the authors that a proportion of actual furfural derivatives, notably hydrodyfurfurals, are present in the lignocellulose, to which, in fact, certain of their characteristic colour reactions are to be ascribed. These, however, are small in amount, and being easily removed, without affecting the essential character of the lignocellulose, may be regarded as products of secondary changes.

In the reaction of the lignocellulose with chlorine it is found that HCl is formed approximately equal in weight to the Cl, combining as lignone chloride. It is to be concluded, therefore, that the reaction is simple and unattended by secondary oxidations of any moment.

The lignone complex when chlorinated, though readily removed from the cellulose, has not been further resolved by any treatments which can be accounted for by quantitative statistics. The evidences as to its constitution are as follows:
(1) As regards the constituent group which combines with chlorine. The lignone chloride when carefully heated gives a sublimate containing chloroquinones; treated with nascent hydrogen it yields trichloropyrogallol; the reaction with sodium sulphite is identical with that of the chlorinated deriva-

tives of pyrogallol, viz. mairogallol and leucogallol. These chlorides in turn have been shown to be derived from oxyquinone groups of the general type—

$$CO \underset{C\!-\!-\!-\!C}{\overset{CH=CH}{\diagup\!\diagdown}} CH_2$$
$$(OH)_2 \; (OH)_2$$

(Hantzsch and Schniter, Berl. Ber. 20, 2023.) the presence of which in the lignone complex consistently accounts for the most characteristic features of the lignocellulose. (2) The residue of the lignine complex, while of similar empirical composition, i.e. approximately, $C_{2n}H_{2n}O_n$, has very different constitutional relationships. Since it readily breaks down under the action of dilute chromic acid in the cold, to acetic acid as a main product, it might be formulated by one of the many alternative $CO\text{-}CH_2$ groupings; and with the keto-R.-hexene groups above the entire complex may be expected to show the constitutional features of the pyrone group.

The further chlorination of the lignone chloride in solution in glacial acetic acid has been studied by the authors (J. Chem. Soc. 1883, 43, 18–21).

The products investigated were obtained from jute (a), and from the fibre (f.v.b.) of the monocotyledonous *Musa paradisiaca* (b). The analysis of the further chlorinated products showed a higher percentage of chlorine (33·75), the results also establishing for both products the empirical formula $C_{38}H_{44}Cl_{11}O_{16}$. The reaction needs further investigation.

The similarity should be noted of the empirical formulæ of the halogenated derivatives of these unsaturated fibre-compounds with those established by Sestini for the so-called sacchulmic compounds (Gazzetta, 1882, 292; J. Chem. Soc. 1882, 1182). These compounds are obtained from the carbohydrates by various processes of dehydration, and, more particularly, spontaneous decompositions or decay of vegetable (cellulosic) matter ('humus').

(2) *Bromine.*—Bromine attacks the lignocelluloses in presence of water; the brominated compound which results

resembles the quinone chloride above described, but the reaction with this halogen is relatively incomplete. After removing the brominated product by hydrolysis with alkaline solutions, and again exposing to bromine water, further reaction of the same kind ensues. Proceeding in this way the lignone constituent is completely removed as alkali soluble derivatives, and cellulose is isolated.

If, on the other hand, the lignocellulose be dissolved in the $ZnCl_2.HCl$ reagent (p. 9) and bromine added, the conditions are more favourable for combination. On precipitating by water, after standing some time, a brominated derivative is obtained, containing 10·2 p.ct. Br (equivalent to 4·5 p.ct. Cl). After standing 16 hours, during which period the cellulose is largely hydrolysed to soluble derivatives, a brominated derivative is obtained, containing 19·5 p.ct. Br (equivalent to 8 p.ct. Cl). Even under these conditions, therefore, the bromine is taken up in considerably less proportion than the chlorine. When the lignone is completely isolated from the cellulose, e.g. by digestion with alkalis at elevated temperatures, it is then brominated in higher proportion. Compounds $C_{17}H_{14}Br_4O_6$, $C_{15}H_{12}Br_4O_5$, $C_{27}H_{28}Br_4O_{10}$ have been obtained from the non-cellulose of esparto, isolated from the alkaline by-products of the papermaker's boiling or pulping process (p. 209; J. Chem. Soc. 41, 94).

(3) *Iodine.*—The lignocelluloses absorb iodine from its aqueous solution and are coloured a deep brown. The reaction has been quantitatively investigated, showing that jute takes up 12·5 p.ct. from decinormal solution in potassium iodide, but the proportion varies according to the concentration of the solution and conditions of the digestion. When these are kept uniform the proportion of the halogen absorbed is constant. The resulting compound, however, is of a loose description, the iodine being easily removed by solvents.

The following experiments with $\frac{n}{10}$ iodine solution in potassium iodide may be cited:

Weight of fibre	Vol. and composition of solution	Absorption p.ct.
2·117	60 c.c. $\frac{n}{10}$I	12·2
2·635	60 ,,	11·3
2·726	60 ,,	13·0
2·463	30 ,,	9·0
2·500	30 ,,	9·8

The absorption, therefore, depends upon the ratio of fibre-substance to iodine solution. This is more clearly shown by the following parallel determinations:

Weight of fibre	Vol. and composition of solution	Absorption p.ct.
2·223	22·2 c.c. $\frac{n}{10}$I + 22 c.c. Aq	6·01
2·374	23·7 ,, + 47·4 ,,	4·8
2·560	25·6 ,, + 76·8 ,,	3·2

It was finally established that, on digesting the fibre-substance at 18° C. with *twenty times its weight of the $\frac{n}{10}$ iodine solution* as ordinarily prepared, *the absorption is constant at* 12·9–13·3 *p.ct.*

It is to be noted that the celluloses also absorb a certain proportion of iodine under similar conditions, viz. 3–4 p.ct. when digested with 20 times their weight of the decinormal solution.

Decompositions of Lignocelluloses, with Resolution into Constituent Groups.—We have already shown that the lignocelluloses are attacked by hydrolytic agents and partially resolved into soluble products. These products, though doubtless of lower molecular weight than the original fibre-substance, preserve its essential characteristics, and the results show that the lignocellulose reacts as a homogeneous compound. When exposed, on the other hand, to the

action of bodies which selectively attack particular groups, its highly complex constitution is brought into evidence.

The reactions with the halogens just described, although reactions of combination, also partake of the character of decompositions, as the evidence has shown. We have now to deal with the decompositions of the lignocellulose in their order, and to emphasise the evidence which they afford as to the relationships of the constituent groups of its complex molecule.

(1) *Non-oxidising acids.*—(*a*) *Hydrochloric acid.*—The fibre-substance boiled with the acid of moderate concentration (1·06 sp.gr.) is profoundly attacked. Furfural distils, and may be quantitatively estimated as already described (p. 99).

The residue is a brownish-black mass of high carbon percentage, presenting some features of resemblance with the original, chiefly in its reactions with chlorine and nitric acid. It is an ill-defined complex, however, and has been only superficially investigated.

(*b*) *Hydriodic acid* acts similarly in the earlier stages of its action. The reaction with this hydracid is made use of in the quantitative estimation of the $O.CH_3$ groups of the lignocellulose. The following determinations have been made in normal specimens:

	(1)	(2)
OCH_3	4·5	4·6 p.ct. of lignocellulose

The acid acts, of course, as a deoxidising acid, and the residue of the reaction is deserving of investigation with the view to determine the limit of deoxidation.

(*c*) *Sulphuric acid.*—The dilute acid at the boiling temperature acts similarly to hydrochloric acid; the volatile products of the decomposition are furfural and acetic acid. In the concentrated acid the lignocellulose dissolves, forming a purple brown solution. On pouring the solution into water a 'condensed' product is precipitated in dark brown flocks; when dried it has the following composition:

C . . . . 64·4
H . . . . 4·4
O . . . . 31·2

On diluting and distilling, acetic acid is obtained. The amount formed in this way is 4-5 p.ct. of the lignocellulose. Acetic acid is therefore a product of hydrolysis of the lignocellulose, which contains a certain proportion of $CH_2.CO.O$ groups.

Nitric acid (dilute), in presence of urea, acts as a non-oxidising acid, and similarly to the above.

Other acids act in similar directions, and in greater or less degree, according to the nature of the acid and the conditions of its action.

Alkalis.—The hydrolysing action of the alkalis in boiling aqueous solution has already been discussed.

At elevated temperatures (150-180°) solutions of the alkaline hydrates (2-3 p.ct. Na_2O) effect a complete resolution of the cellulose and lignone, the latter being obtained in solution in the form of acid derivatives. In addition to acetic acid the solution contains acids of high carbon percentage, which are precipitated on adding mineral acids to the alkaline solution. These bodies have been investigated by Lange (Zeitschr. Physiol. Chem. 14, 217), but as the products described by him were obtained from lignocelluloses of another group—viz. the woods—and under more severe conditions of treatment, they will be dealt with subsequently. Jute, however, yields very similar products, viz. acid bodies of high carbon percentage (60-61), giving Cl substitution derivatives. The cellulose retains residues of these bodies, but they are easily eliminated by treatment with hypochlorite solution. The cellulose is then obtained as a white pulp, consisting of the disintegrated fibre-elements or ultimate fibres. The yield from normal specimens is about 60 p.ct. Only the more resistant cellulose α survives the treatment, the cellulose β, together with the lignone complex, being converted into soluble derivatives.

Extreme action of alkaline hydrates.—With the caustic alkalis in concentrated solution and at temperatures exceeding 120°, much more drastic decompositions take place, the entire molecule being attacked. For complete resolution into simple molecules (oxalic, acetic, and carbonic acids) the proportion of alkaline hydrate to lignocellulose requires to be 2-3 to 1, and the temperature raised to 250°, and maintained at that point for some hours. Thus, on heating jute for 8 hours at 250° with 3 times its weight of KOH, the yields of the main products were: acetic acid, 37·0 p.ct.; oxalic acid, 53·3 p.ct. of the lignocellulose (J. Soc. Chem. Ind. 11, 966). The action is an oxidising action, in the sense that hydrogen is expelled; gaseous carbon compounds (CO, CH_4) are formed in relatively small quantities.

DECOMPOSITIONS BY OXIDANTS.—(1) *Acid.*—Certain of these are more important as contributing to the elucidation of constitutional points.

(a) *Chromic acid.*—The direction of attack of this oxidant depends upon the auxiliary conditions, chiefly upon the presence of hydrolysing acids. With the CrO_3 alone, the interaction with the lignocellulose is at first one of simple combination; afterwards the CrO_3 fixed is gradually deoxidised. Under these circumstances the lignocellulose suffers a very slight loss of weight. In presence of acids, however, the fibre-substance loses in weight, and the insoluble residue is affected more or less. The following results may be cited in illustration:

CrO_3 *alone.*—(1) 4·5 grms. jute, containing 0·7 p.ct. ash constituents, digested 18 hours in 1 p.ct. solution CrO_3

(250 c.c.) at 15-18°. Weight of product, 4·55 grms.; with ash, 5·0 p.ct.; CrO_3 fixed, equivalent to 4·2 p.ct. Cr_2O_3.

(2) 1·8 grm. fibre; 100 c.c. 1 p.ct. CrO_3; 16 hours at 15°. Product, 1·82 grm.; ash, 4·3; CrO_3 fixed, equivalent to 3·7 p.ct. Cr_2O_3.

CrO_3 and acetic acid.—1·8 grm. fibre; 100 c.c. 1 p.ct. CrO_3 containing 4 grms. acetic acid: (*a*) digested 16 hours, (*b*) digested 20 hours.

Product	P.ct. of original	Ash p.ct.	C p.ct. in product
(*a*) 1·70	94·4	1·8	43·2 42·8
(*b*) 1·65	91·7	1·8	42·0

The products were largely soluble in dilute alkaline solutions; the lignocellulose reactions were faint; the characteristics of the products were those of the oxycelluloses.

CrO_3 and sulphuric acid (dilute).—(1) 0·9 grm. fibre; 100 c.c. solution containing 0·874 CrO_3 and H_2SO_4: (*a*) 1·6 grm., (*b*) 3·2 grms., (*c*) 4·8 grms.

Product	P.ct. of original	CrO_3 consumed	—
(*a*) 0·840	93·3	0·510	Traces only of gaseous products evolved
(*b*) 0·801	89·0	0·470	
(*c*) 0·768	85·3	0·430	

The solution of the fibre-substance increases with the increase of hydrolysing acid; the deoxidation of the CrO_3 slightly decreasing.

(2) Jute, 0·9 grm.; 100 c.c. solution containing variable quantities of CrO_3 and H_2SO_4 as under (CrO_3 solution, 0·842 CrO_3 per 10 c.c.; $H_2SO_4 = 7·59$ per 10 c.c.):

—	(1)	(2)	(3)	(4)
CrO_3 solution	10 c.c.	15	20	25
H_2SO_4 solution	10 c.c.	15	20	25
Yield of oxycellulose	72·8 p.ct.	65·0	53·3	45·0

The oxycelluloses obtained were soluble in alkaline solutions and in nitric acid (1·43 sp.gr.). Under these more severe conditions there is an increasing evolution of gas, and from (4) 75 c.c. were collected.

In the process of oxidising with chromic acid in presence of a hydrolysing acid (H_2SO_4), acetic acid is formed. Oxidised by its own weight of CrO_3 in presence of excess of normal sulphuric acid, the fibre-substance yields from 12-13 p.ct. $C_2H_4O_2$.

It is evident that chromic acid oxidations of the fibre-substance can be controlled within any prescribed limits. From the investigations, of which the above are typical series of experiments, it was concluded—

(1) That the keto R. hexene groups yield most readily to the action, and may in fact be selectively attacked and eliminated; (2) that with a net loss of weight of 10 p.ct. the lignocellulose is converted into an oxycellulose containing 42-43 p.ct. carbon, and yielding the same percentage of furfural (HCl distillation) as the original fibre. The furfural-yielding complex is not, therefore, radically affected by the treatment. (3) As the amount of oxygen expended (CrO_3 deoxidised) is relatively small—approximately 1 mol. per unit weight $C_{12}H_{18}O_9$ of lignocellulose—and would appear to be chiefly consumed in oxidising the portion passing into solution, the relatively large reduction in *carbon percentage* of the insoluble residue is due to simultaneous fixation of *water*. (4) It appears, in fact, that the furfural-yielding complex is by such action converted into an oxycellulose.

Chromic Acid and Sulphuric Acid (Conc.).—When the lignocelluloses are dissolved in concentrated sulphuric acid, the addition of chromic acid determines complete combustion of the carbon to gaseous products CO_2 and CO. The proportion of CO formed is usually very small. As both gases, however, have the same molecular volume, a determination of the total gas evolved gives by calculation the carbon contents of the substance. The method is available for analytical purposes, and will be found fully described in J. Chem. Soc. 53, 889.

As 1 mgr. of lignocellulose gives approximately 0·9 c.c. CO_2 under ordinary conditions, it will be seen that trustworthy results can be obtained with very small quantities of substance; and as the entire operation takes only a very few minutes, the method is extremely useful for rapid approximate analyses of products obtained in the course of investigation.

(*b*) *Nitric acid.*—In the interaction of nitric acid with the fibre-substance, in presence of sulphuric acid, it has been already shown that decomposition (oxidation) supervenes after a few minutes' exposure. The acid (1·5 sp.gr.) alone attacks the lignocellulose still more rapidly and energetically; as the oxidation is of a 'wholesale' character, its investigation would not throw much light upon the constitution of the fibre-substance. The acid of 1·43 sp.gr. acts more gradually; there is direct combination in the first instance attended by deoxidation. A yellow product is obtained differing but little in weight from the original, and containing 2·0–2·5 p.ct. N. With the progress of the oxidation there is considerable disintegration of the fibre-substance and conversion into soluble derivatives, but of an ill-defined character. With the dilute acid, on the other hand, a very gradual resolution ensues, and the reaction has been carefully investigated. The main results determined are these: the lignocellulose is entirely resolved into insoluble cellulose (*a*) and soluble derivatives of the remaining groups, with a proportion of acid (HNO_3) equal to 25 p.ct. of the fibre-substance. The specific action of the acid takes place at any dilution not exceeding 30 Aq : 1HNO_3 (by weight), and any temperature within the range 40–100°. The most convenient conditions are with the acid at 7–10 p.ct. HNO_3 and temperature 60–80°. Under these conditions there is considerable evolution of gas, and of very complex composition.

The course of the reaction may be thus described. The

lignocellulose is changed in colour to a bright yellow which gradually changes to lemon yellow, and after some hours' digestion, to white. If the digestion be interrupted at the yellow stage, the fibre washed and digested with boiling alcohol, a bright yellow solution is obtained; and on driving off the alcohol a gummy body is left, characterised by great instability, reducing Fehling's solution in the cold, yielding furfural on boiling with acids, and progressively decomposed on heating at 100° (in presence of water) with evolution of gaseous products. The substance retains from 1-2 p.ct. N, but in a very unstable form, being entirely split off on heating with water. This ill-defined product we may term, for obvious reasons, the *intermediate body*. These results will be appreciated from the following statement of the final products of the decomposition.

Lignocellulose and Dilute Nitric Acid.

Solid products: Cellulose α 63-66 p.ct.   Oxalic acid 4·0-5·5 p.ct.   Intermediate body, 5·3-5·8 p.ct.
Volatile acid: Acetic acid, 14-18 p.ct.
Gaseous products: From HNO_3 From fibre-substance
$N_2O_4, N_2O_3, N_2O, N_2, HCN$ CO_2, CO, HCN
(Representing about 50 p.ct. of the N of the HNO_3)

The most notable features of the decomposition are—

(1) As regards the HNO_3—(a) The reaction depends upon the presence of nitrous acid; the addition of urea entirely arrests the specific action of the acid, and it then behaves exactly as the non-oxidising mineral acids. (b) The direct deoxidation of nitric acid never proceeds beyond the formation of NO; the presence of N_2O indicates the formation of a hydroxime, and its decomposition by further reaction with nitrous acid. The formation of HCN also appears to result from the dehydration of a product of this nature, and this conclusion is confirmed by the observation that HCN appears in

greatest quantity at the end of the reaction and as the temperature is raised to 100°. (c) The presence of N_2 indicates a still further deoxidation—or hydrogenisation of the N to ammonia.

(2) As regards the fibre-substance, the keto R. hexene groups are rapidly oxidised, and entirely broken down. No 'aromatic' products are formed, and the result is in perfect accord with the general view we have taken of their constitution.

The *furfural-yielding* complex and the *β-cellulose* group are more gradually resolved, and both probably contribute to the large yield of acetic acid. It is obvious that the constitution of both groups is of a special type, unlike the normal grouping of the carbohydrates.

The reaction has been also studied in connection with another group of the lignocelluloses—viz. the woods—and will be again referred to (p. 212).

Joint action of oxides of nitrogen and chlorine.—F. Schulze's method of eliminating the non-cellulose groups of the lignocelluloses, in the isolation and estimation of cellulose, has been already described. It consists in a prolonged digestion in the cold with nitric acid (1·1 sp.gr.), with addition of a small proportion of potassium chlorate. The reaction has not been investigated in regard to the by-products. The mechanism of the decomposition will be evident from what has been stated in regard to the actions of chlorine and of nitric acid upon the lignocellulose.

(?) ALKALINE OXIDANTS.—The actions of this group of reagents are of considerable technical importance, as upon them depend the various bleaching methods in common practice; but they have not been sufficiently investigated to throw light on theoretical points.

Potassium permanganate acts, of course, as an oxidising agent pure and simple. The limit of deoxidation in basic or neutral solution is the oxide MnO_2, which is deposited upon, and in intimate combination with, the lignocellulose. If removed by treatment with sulphurous acid it does not further attack the lignocellulose, the treatment merely revealing the bleaching action accomplished in this stage of the deoxidation. By treatment with sulphuric acid the lignocellulose undergoes further oxidation, and with hydrochloric acid chlorine is liberated and combines with the fibre-substance. These reactions are, however, of little importance. The *permanganate bleach* is too costly for general adoption in the case of jute fabrics. From its simplicity, it is a useful treatment in the laboratory, whether for removing coloured impurities from the raw fibre, or from cellulosic products separated by any of the processes already described.

Hypochlorites.—Bleaching powder solution (calcium hypochlorite) and the equivalent sodium compound act, in presence of excess of the base, as oxidising compounds; but as by oxidation and attendant hydrolysis, acid derivatives are formed from the lignocellulose, the use of a 'neutral' solution of the bleaching solution often leads to chlorination of the fibre-substances, owing to liberation of hypochlorous acid. Neglect of this probability has led to disastrous results in the bleaching of jute piece goods; and a full discussion of the matter, in both practical and theoretical bearings, will be found in the Bull. Soc. Ind. Mulhouse, 1880.

The danger is avoided by ensuring the presence of excess of base; this is more easily controlled in solutions of the soda compound, which are therefore to be preferred. After bleaching with the hypochlorites, the fibre or fabric should be well washed, and plunged for a short time into sulphurous acid solution which removes the last traces of oxidising compounds. After again washing, the lignocellulose may be dried without change of colour.

Hypobromites.—The hypobromites of the alkalis attack the lignocelluloses profoundly; amongst the final products of decomposition bromoform and carbon tetrabromide are obtained in some quantity. (Compare N. Collie, J. Chem. Soc. 1894, 262, which contains the results of a general investigation of the reaction.)

In regard to the alkaline oxidants it may be said generally, in conclusion, that they attack the non-cellulose constituents of the lignocelluloses in greater degree, but their action extends to the cellulose also. They are therefore of little present use as 'pioneer reagents,' and have moreover secured no systematic investigation.

Other Decompositions of Lignocellulose.—There are a number of decompositions remaining to be described which do not fall within any group classification : these will now be dealt with in the order of their importance.

Interaction of lignocellulose with sulphites and bisulphites.— Jute, when heated at high temperatures with solutions of the alkaline sulphites, or of the bisulphites of the alkaline earths, is directly resolved into cellulose (insoluble) and soluble compounds of the lignone complex with the sulphites. A similar treatment of pine-wood, attended by the same results, is the basis of the now highly developed 'sulphite wood pulp' industry. The process and the reactions upon which it is based will be described in detail in a later section ; and as the general principles apply to the lignocellulose with which we are dealing, it is unnecessary to anticipate the fuller treatment of the subject.

The theory of the reaction is deducible from the following considerations : The lignocellulose when heated with water only, at high temperatures ($140-160°$), is profoundly attacked ; a considerable proportion of the fibre-substance passes into solution (hydrolysis), and the residue of disintegrated fibre resembles the product obtained by digestion with dehydrating acids. It is obvious, *à priori*, that a reaction of this kind will proceed to the limit representing equilibrium between the hydrolysing and condensing influences. If, now, a substance be present capable of uniting with the products of hydrolysis in such a way as to prevent them entering further into reaction, the resolution will proceed without secondary complications to the limit determined by the constitution of the lignocellulose. *Sodium sulphite* is a reagent fulfilling these conditions : acid products combine with the base, and aldehydic products with the bisulphite residue. In this case, however, the hydrolysis being thrown chiefly upon the water, a high temperature ($160°$) is required to effect complete decomposition. By substituting bisulphites, the hydrolysis is aided from the first by sulphurous acid, and the decomposition is completed at lower temperatures (130–$140°$). That the hydrolysing action of sulphurous acid is a powerful factor is evident from the fact that an aqueous solution of this acid, containing 7–8 p.ct. SO_2 (which, of course, requires to be prepared under pressure), will itself resolve the lignocellulose at the lower temperature of 95–$105°$. The reaction with the bisulphites is, however, in many respects simpler, and complete decomposition is effected with solutions containing 3–4 p.ct. SO_2.

The yield of cellulose is 63–66 p.ct. of the lignocellulose, and is composed therefore of the more resistant α cellulose ; the β-cellulose is hydrolysed under these conditions also, and passes into solution with the lignone. The soluble derivatives preserve the features of the original lignone ; combining with the halogens to form substitution products, and yielding furfural on boiling with hydrochloric acid. All the reactions of the product indicate that it is a *sulphonated* derivative of the lignone complex of the original fibre. For the further discussion of the reaction see p. 198.

Compound Celluloses

The following observations upon the behaviour of the fibre when treated with water at high temperatures may be cited. The experiments were conducted in glass tubes.

(1) Heated 12 hours at 110° C.: Slightly attacked. Loss in weight, 11·0 p.ct.

(2) Heated 10 hours at 120-130°: Only slightly attacked. Heated further 10 hours: Fibre disintegrated. Loss of weight, 27·5 p.ct. Solution contained furfural.

(3) Heated 9 hours at 140°: Completely disintegrated. Loss of weight, 22·6 p.ct.

Analysis of Disintegrated Fibre.

C 48·30 p.ct. } Yield of cellulose (Cl method), 76·8 p.ct.;
H 5·16 p.ct. } calculated on original fibre, 59·5 p.ct.

(4) Heated with water and barium carbonate 9 hours at 140°: Colour changed to brown. Fibre *not* disintegrated. Loss of weight, 20·0 p.ct. Product yielded: cellulose, 79·3 p.ct.; calculated on original, 63·5 p.ct.

(5) Heated with solution sodium sulphite (5 p.ct.) 10 hours at 120-130°: Loss of weight, 19·0 p.ct. Fibre disintegrated; fibre and solution *colourless*. Cellulose p.ct. on product, 84·6; p.ct. on original fibre, 68·7.

(6) Heated with sodium bisulphite solution (2·6 p.ct. SO_2) 10 hours at 115°: Fibre disintegrated; fibre and solution colourless. Loss of weight, 19 p.ct. Cellulose p.ct. on product, 73·5; p.ct. on original fibre, 65·4.

Animal digestion of the lignocelluloses.—The urine of the herbivora contains hippuric acid as a characteristic constituent. The origin of this compound, and more particularly its benzoyl group, has been the subject of considerable discussion and controversy, but the evidence points unmistakably to the lignocelluloses of the various fodders as the source of the product. It appears from what we now know of the constitution of the lignone complex, that its R. hexene and $CO-CH_2$ groups may, without unduly straining the probabilities, be regarded as undergoing transformation, in the processes of animal metabolism, to the compound in question. The problem has been specially investigated by Meissner and Sheppard (1866), Stutzer (Berl. Ber. 8, 575), Weiske (Ztschr. Biol. 12, 24).

Spontaneous decomposition of the lignocelluloses.—Jute is sometimes baled in a damp state, or wetted by sea water in course of shipment, and the fibre in the interior of such bales is found to undergo considerable chemical change, attended by structural disintegration. A specimen of the fibre thus disintegrated was found to present the following features:

Soluble in water	10·0 p.ct.	
Soluble in 1 p.ct. NaOH	23·0 ,,	
Cellulose	60·4	58·8

The aqueous exhaust of the fibre was astringent to the taste, was precipitated by gelatin solution, and gave coloured reactions with iron salts. It was digested on barium carbonate, filtered, evaporated, and the residue resolved by alcohol into (1) a soluble body of 'neutral' characteristics, which on analysis gave numbers expressed by the empirical formula $C_{26}H_{34}O_{16}$. This substance, on fusion with potash, gave some phloroglucol and a large yield of protocatechuic acid. (2) An insoluble body, the Ba salt of an acid, which on analysis gave numbers expressed by the formula $BaC_{29}H_{42}O_{29}$.

The investigation of these products dates from 1880 (J. Chem. Soc. 41, 93), and they were not examined for determination of the now well-established 'constants' (p. 157). From the above results, only the main features of this spontaneous decomposition of the lignocellulose are evident, viz. a resolution of the lignone complex into more and less oxidised groups; the latter representing transition to aromatic products of definite and ascertained relationships, the former having features in common with the group of pectic compounds. But they have the additional interest of suggesting, in a very direct way, the origin of the astringent substances or tannins, widely distributed throughout the plant world; and not only of the tannins, but more generally derivatives of the trihydric phenols. This problem is of the greatest interest both from the chemical and physiological standpoints. It involves varied transitions from aliphatic to cyclic compounds, and a prominent feature of the synthetic activity of the plant, the elucidation of which is the immediate objective of organic chemistry.

Of the endless variety of excreted products in vegetable growth —e.g. essential oils, waxes, alkaloids, and 'aromatic' products— the *tannins* have specially attracted the attention of physiologists. An important monograph on the subject has recently appeared: Grundlinien zu einer Physiologie des Gerbstoffs, by G. Kraus, Leipzig, 1888. The work contains the results of extensive experimental investigation of the origin, distribution, fate, and function of tannins in normal growths. A *résumé* of the evidence shows generally:

(1) That the tannins are formed in leaves under the same conditions as are necessary for general assimilation, but is an independent process. The tannins thus formed are transmitted through the leaf stalk, and distributed through the permanent structures.

(2) They are also formed in processes of growth in the dark, e.g. growth of rhizomes, unfolding of buds, &c.

(3) Also in isolated cells and tissues, where they remain.

(4) The tannins take no further direct part in plant assimilation; they are end-products.

In dealing with the question of the sudden increase observed in passing from the sap to the heart wood of many trees, the author speaks as follows: 'The only satisfactory explanation would be the assumption that the tannin in this case is formed locally, i.e. in the wood tissue itself, the parent substance being the tissue of the medullary rays and wood-parenchyma. As to the possible mechanism of such a process, however, we are in total darkness.' That, we venture to think, is no longer the case.

Destructive Distillation.—The decomposition of jute by destructive distillation has been specially investigated by

Chorley and Ramsay, who obtained the following results (J. Soc. Chem. Ind. 11).

Weight of fibre	71 grms.	73 grms.
	P.ct.	P.ct.
Charcoal	28·71	32·87
Total distillate	57·70	43·15
Carbonic anhydride	—	12·33
Other gases (and loss)	—	11·65
		100·00
Volume of gas	3,000 c.c.	2,500 c.c.
,, ,, per 100 grms.	4,220 ,,	3,420 ,,

Composition of Products p.ct.

		P.ct. vol.	
Gas	Carbon monoxide	78·80	85·29
	Oxygen	3·01	1·73
	Residual gas	18·19	12·98
		P.ct. fibre	
Distillate	Tar	14·78	6·85
	Acetic acid	0·40	1·40
	Methyl alcohol	—	10·08

The chief features to be noted in the products are the low yields of charcoal (compare p. 69) and acetic acid, and the high yields of carbonic oxide and methyl spirit. The thermal features of the decomposition are remarkable : heated gradually to 320° the temperatures within the distilling flask and external to it follow the ordinary course ; but at 320° (external) the temperature within the flask rushes up to 375°, the change being marked by a much increased evolution of gas.

The destruction of a complex substance such as the ligno-cellulose, by heat, involves a highly complicated web of reactions, which it would be impossible to disentangle in detail and in such a way as to throw light on the fate of particular groups. In the main there are, of course, the two opposing factors at work—*dissociation*, giving products of lesser, to those of the least molecular weight (gases) ; and *condensation*, giving products of greater complexity, up to those of indefinite molecular weight (charcoal or pseudo-carbon). More definite features of the condensation are the closing of the C_4O ring (furfural) and the further condensation of the hexene to benzene rings. But to study these and other changes in reference to the parent molecule, it would be necessary to carry out an elaborate series of quantitative observations, varying not only the physical conditions of the distillation (temperature, time, &c.), but the chemical factors by the admixture with the fibre-substance of reagents of known function. Until we have such results the imagination is free to go to work upon such slender materials as are available.

General Conclusions as to the Composition and Constitution of Jute Lignocellulose.—Having thus set forth the general chemistry of the typical lignocellulose, it is important to select and bring together those facts which bear more particularly upon the problem of its constitution. This problem, it may be remarked, cannot be divorced from its essentially physiological aspects : a plant is an assemblage not merely of products, but of processes ; and in investigating a plant tissue, we have not merely to ascertain the quantitative relationships of its constituents, but from the point of view of physiology or organic function to distinguish between organic and excreted products ; further, to endeavour to arrive at their genetic relationships. The history of every tissue is one of continuous modification, and the excreta of plants are, in many cases, the last links of a long chain of transformations. Where such compounds are formed as by-products of the assimilative processes, we cannot as yet hope to have any definite clue to their origin ; but where they originate independently, either by intrinsic or extrinsic modification (e.g. oxidation) of a tissue-substance, the clue may be expected to be found in the constitution of the tissue-substance itself. Or, to put it in another way, if we find associated with a tissue an excreted product of general constitutional resemblance thereto, we cannot avoid the suggestion of genetic relationship. The suggestion becomes an hypothesis upon which investigation can proceed. The lignocelluloses, for instance, afford many indications of such relationship to the tannins. In the jute fibre, tannins are always present in small quantity; the characteristic R. hexene groups of the lignocellulose occupy a definite and close relationship to the trihydric phenol, pyrogallol, to which many of the tannins stand in direct constitutional relationship; and we have described an instance of 'spontaneous' transformation of the lignocellulose into a substance having the essential characteristics of the tannin group. The general discussion of this question belongs to a later section of our subject (see p. 179). It is introduced here in order to show that we cannot attempt to formulate a molecule of a lignocellulose on the lines of a carbon compound of ascertainable molecular weight and such relationships of its constituent groups as are sharply defined and verifiable by synthesis. It is true that when attacked in detail the lignocellulose is resolved into well-differentiated groups, which may be regarded with reservation as constituents of the parent molecule; the reservation being that unless and until the lines of cleavage are proved to be invariable, we cannot consolidate the results of various directions of resolution into a homogeneous view of the parent substance.

We will now point out how far the problem is solved by the evidence available.

(1) THE LIGNOCELLULOSE A HOMOGENEOUS COMPOUND RATHER THAN A MIXTURE.—The evidence for this conclusion is as follows—(a) *Physiological* : general uniformity in composition and reactions; does not vary with age of fibre (i.e. from root

upwards) nor with thickening of cell wall (incrustation) ; preserves essential features through wide range of differences in empirical composition, resulting from differences in conditions of growth. (*b*) General resistance to *resolution* (into proximate constituents), by the action of solvents and hydrolytic agents generally. (*c*) Behaviour in synthetic reactions, chiefly in ferric ferricyanide reaction and formation of nitrates ; resistance of molecule to *resolution.*

(2) GENERAL CHARACTER OF LIGNOCELLULOSE CONSIDERED AS A WHOLE.—The alcoholic characteristics of the lignocellulose are inferior to those of cellulose : the reactive OH groups are fewer in proportion ; CO groups of aldehydic, ketonic, and acid function are present in union, more or less, with the more basic OH groups. The characteristic reactions of the compound (lignone group) are those of unsaturated compounds, and it is, by comparison with the celluloses, greedy of oxygen.

(3) CONSTANTS OF THE FIBRE IN REACTION.—In this connection we refer only to such reactions as throw light upon the relationships of constituent groups, and therefore reactions of decomposition.

Cellulose.—Cl method : average yield, 75·0 p.ct. ; raised, by minimising conditions tending to hydrolysis and oxidation, to 78-82 p.ct.

Br method : average, 72·0 p.ct. ; may be raised similarly to 74-76 p.ct.

Nitric acid (dilute) method, 63-66 p.ct. Alkali method, 56-60 p.ct. Bisulphite method, 60-63 p.ct.

The cellulose is a variable, the variations being due to greater or less hydrolysis. The lignocellulose contains a cellulose of resistant characteristics and a cellulosic constituent which is either isolated as cellulose or dissolved with the lignone complex according to the treatment. This latter cellulose, when isolated, contains $O.CH_3$ groups. The whole cellulose complex gives the following results on analysis:

			Calc. $C_{18}H_{32}O_{16}$
Ultimate analysis . $\begin{cases} C \\ H \end{cases}$	43·0	. . .	42·8
	6·1	. . .	6·3

Proximate . . $O.CH_3$ 1·2   Furfural . 6–8 p.ct.

The cellulose is a hydrate, and a mixture of two celluloses: the β-cellulose contains the methoxyl groups, and gives furfural with condensing acids.

*Lignone.—Chlorination.—*Cl combining with lignone, 8·0 p.ct.; Cl combining as HCl, 8·0 p.ct., calculated on the lignocellulose. Composition of lignone chloride, $C_{19}H_{18}Cl_4O_9$ (containing 26·7 p.ct. Cl), from which we may assume $C_{19}H_{22}O_9$ as the approximate formula for the lignone complex.

From these statistics, and on the assumption that there are no hydration changes of any moment, we may calculate the lignone complex to constitute a little over 20 p.ct. of the lignocellulose.

Continuing this statistical and approximate method of investigation and calculating to carbon percentages—

Cellulose (anhydride), 44·4; lignone, 57·8.
80 × 44·4 ÷ 100 = 38·52
20 × 57·8 ÷ 100 = 11·56

47·08 p.ct. C in lignocellulose.

These results confirm the evidence of the 'quantitative' character of the chlorine reaction.

From a study of the ester reactions of the lignone chloride and of the original lignone group, it is to be concluded that the former contains not more than one alcoholic OH group, even when isolated by treatment of the chlorinated fibre with sodium sulphite solution. Both in union and under resolution, therefore, the two main complexes preserve their general character of complex anhydrides. The union resists the action of all simple hydrolytic treatment, but yields at once when the condition of oxidation or the specific attack of negative groups (NO_2, Cl_2O) is superadded.

It has been shown by the quantitative statistical stu ly of these several reactions, and by the composition of the products, that each has its characteristic line of reso ution or cleavage of the original lignocellulose complex. While the lines of separation of the α-cellulose and β-cellulose residues are well marked, it is doubtful whether the β-cellulose is as sharply separated from the lignone complex. Under the chromic acid treatment it certainly appears that a portion of the latter is converted into a cellulose (oxycellulose), as a result of attendant hydration; moreover the general features of the lignone are largely those of products obtained by the action of condensing agents upon the carbohydrates, and it is not improbable that the general configuration might be so retained that combination with water would restore the carbohydrate, i.e. cellulose, character. Again, if the β-cellulose is a keto-cellulose, as it appears to be, it may exist in a condensed form in the lignocellulose, and thus have features as much in common with the lignone as with the α-cellulose.

These considerations justify the use of the group-term lignocellulose, and at the same time show that the constituent groups lignone-cellulose must not be too easily regarded as fixed quantities. It leaves open the question as to whether the lignone is not genetically connected with the cellulose. This is an important physiological probability which will be met with again in considering the chemistry of the woods, i.e. the lignocellulose of perennials.

Furfural.

	P.ct. of lignocellulose.
Yield from original fibre-substance	8–9
Yield after chlorination	8–9
	P.ct. of products.
Yield after CrO$_3$ treatment	8–9
Yield from isolated cellulose Cl method	7–8

The origin of this characteristic product of decomposition is localised mainly in the cellulose complex, the group from which it is derived being isolated as a cellulose (β-cellulose) by the chlorination method. The lignocelluloses in their 'natural' condition appear also to contain hydroxyfurfurals in small proportion, and to which their characteristic colour reactions with phenols, especially phloroglucinol, are probably due.

Methoxyl.—The presence of O.CH$_3$ groups is another characteristic feature of lignification, i.e. of a lignocellulose. In jute the total yield is 4·6 p.ct.; the major proportion of the methoxyl is localised in the lignone complex. A certain proportion appears in the cellulose isolated by the chlorination process, which is further suggestive of the relationships of the lignone to the β-cellulose previously discussed (p. 159).

Assuming that the whole of the O.CH$_3$ is contained in the lignone complex, the empirical formula assigned to this may be calculated to contain two such groups, and would become C$_{17}$H$_{16}$O$_7$.2OCH$_3$, a formula similar to that arrived at for a product obtained from the lignone of coniferous woods (p. 201), viz. C$_{24}$H$_{24}$O$_8$.(OCH$_3$)$_2$.

Acetic acid is an important product of resolution of the lignocelluloses by the action of hydrolytic and oxidising agents, and under conditions of very limited intensity. The source of this product is the lignone complex, and when this is broken down by treatment with chromic acid (in presence of sulphuric acid) the yield amounts to 50–70 p.ct. of its weight. The conditions of its formation point to its being a product of hydration rather than oxidation; it is probable that more complex ketonic acids are first produced, and further resolved on distillation, especially in presence of excess of the oxidant.

It appears from this that the lignone complex contains, associated with the oxyquinone groups, a large proportion of $CO.CH_2$ groups, the configuration of which remains undetermined. It is probable that groups allied to dehydracetic acid are represented, and a pyrone grouping of a portion of the complex would account for the production of acetone as a first product of destructive distillation (pp. 154-206).

(2) **Other Types of 'Annual' Lignocelluloses.**—The chemistry of the jute fibre might be presumed to cover the essential features of the lignification of bast fibres generally; so far as investigation has gone, this appears indeed to be the case. This statement, of course, must not be taken as suggesting *identity* of constitutional features. Comparative investigation of the bast tissues of the dicotyledonous annuals generally has not as yet been attempted. Such work is called for, and it is impossible to predict the influence which the results might have in extending our grasp of the physiology of the exogenous stem. Of those which are lignocelluloses, jute is undoubtedly typical, and the methods adopted for this fibre may be extended to the group.

The process of lignification, however, is by no means limited to particular tissues, and we have now to deal with other representative cases of the formation of lignocelluloses in 'annual' structures.

'GLYCODRUPOSE.'—*The hard concretions of the flesh of the pear* are composed of a lignocellulose giving the typical reactions of the jute fibre. This product was investigated some years ago by Erdmann (Annalen, 138, 9). The concretions are isolated from the parenchyma of the fruit in which they are imbedded, by long boiling with water, rubbing down to a pulp, washing away tne cellular débris, and thus, by continued mechanical action and washing, entirely freeing them from the matrix of softer tissue. The substance of these concretions gives constant results on elementary analysis, expressed by the empirical formula $C_{24}H_{36}O_{16}$; to this complex Erdmann gave the name *Glycodrupose*, and he regards it as resolved on boiling with hydrochloric acid according to the equation :

$$\underset{\text{Glycodrupose}}{C_{24}H_{36}O_{16}} + 4H_2O = \underset{\text{Glucose}}{2C_6H_{12}O_6} + \underset{\text{Drupose}}{C_{12}H_{20}O_8}.$$

Drupose, on fusion with potash (KOH), yields aromatic products amongst which pyrocatechin was identified. Glycodrupose, on boiling with dilute nitric acid, gives a residue of pure cellulose. These results were repeated by Bente (Berl. Ber. 8, 476), and in general terms confirmed, though the analytical results varied somewhat from the above.

From our present point of view, the interpretation of these results by these investigators is open to question in more than one direction, but they certainly establish the following points :

The concretions represent a compound cellulose, of which the non-cellulose is easily converted into aromatic derivatives. This compound cellulose gives constant results when analysed, whether for its elementary or proximate constituents, and is therefore a chemical individual. The authors, on the other hand, in investigating the product some years ago, noted a very close resemblance in all the reactions of this complex with those of jute. It may therefore be included amongst the *Lignocelluloses*. It may be also noted that the formula assigned to the complex by Erdmann differs by only one O atom from the empirical formula which we have used for the jute substance :

$$\underset{\text{Glycodrupose}}{C_{12}H_{18}O_8} \qquad \underset{\text{Jute lignocellulose}}{C_{12}H_{18}O_9}$$

The authors have made (1883) the following determinations of the constituents of these concretions:

Inorganic constituents (ash)	0·91	
Cellulose	26·0	34·2
[1] Furfural	18·0 p.ct.	
Loss on boiling in 12 p.ct. HCl (30 mins.)	53·6	

In conclusion, we can only call attention to the desirability of re-investigating the product, and, upon the evidence of close similarity to the typical lignocellulose, of adopting, at the outset, the general plan of investigation laid down for such compounds.

The formation of a lignocellulose under such totally different conditions from those which obtain in a flowering stem is of especial significance in regard to the physiology of the production of such compounds.

THE LIGNOCELLULOSES OF CEREALS.—Both in the straws of cereals, and the seed envelopes of the grain, there is a typical and characteristic process of lignification. With the formation of quinone-like bodies, as in jute, there is associated the production in the tissue of a large quantity of pentosan derivatives.

The composition of *brewers' grains* has been carefully investigated by Schulze and Tollens (Landw. Vers.-Stat. 40, 367), and an abstract of their results is given in Section III. p. 259. From the more recent results of Tollens this material has been found to yield 16·03 p.ct. furfural, corresponding to 26·93 p.ct. of pentosan. A considerable proportion of the pentosan constituents may be directly hydrolysed to pentaglucose; on the other hand, a not inconsiderable proportion is so intimately united to the cellulose as to resist hydrolytic treatments of some severity. The lignone constituent was not specially

---

[1] The furfural was estimated by the colorimetric method of comparison with a standard solution of furfural (V. Meyer, Berl. Ber. 11, 1870), the only method available at the time. As a specimen of jute similarly investigated gave 10·6 p.ct. furfural, it is probable that the above determination is 2-3 p.ct. in excess of the true number.

investigated by Tollens in regard to its more characteristic groups, the researches being chiefly directed to the furfural-yielding groups. What we have to emphasise is the recognition by Tollens that in this tissue-substance the various groups are so united as to constitute a homogeneous complex. This tissue has the closest resemblance to the grain-bearing *straws*, which have been recently investigated by C. Smith and the authors (J. Chem. Soc. 1894, 472 ; Berl. Ber. 1894, 1061).

The starting-point of these researches was the observation, already noted (p. 84), that the celluloses, isolated from their stem tissues, themselves give a large yield of furfural when boiled with hydrochloric acid ; at the same time *none of the reactions of the pentaglucoses*. It appears from these researches, and from subsequent results, that from germination, continuously with the growth of the stem, there is a steady increase in the proportion of furfural-yielding constituents, and that these are mainly utilised in building up the *permanent tissue* of the stem. These results are noted *pari passu* with *lignification*, and they further generalise the chemical features of the process which were brought out in connection with the jute fibre—viz. (1) the cellulose of a lignified tissue is, when isolated, found to be invariably an oxidised and furfural-yielding cellulose ; (2) in the non-cellulose, pentosan groups are present in association with an easily hydrolysable oxycellulose, *and* with unsaturated or keto R. hexene groups.

As lignocelluloses, the straws are generally differentiated from the typical lignocellulose, (1) by their structural complexity ; (2) by their lower carbon and proportionately greater oxygen percentage ; (3) by the relative susceptibility of the non-cellulose to hydrolysis ; (4) by the much lower percentage of cellulose and the composition of this cellulose.

As a consequence of these differences, the straws are more easily attacked by the thiocarbonate treatment. The following are the results of an experiment carried out under the usual conditions.

Undissolved by treatment	40·3 p.ct
Soluble and reprecipitated by acids	32·4 ,,
,, and not reprecipitated by acids	27·3 ,,

The straws and products of this class have thus been investigated in various directions, but by no means exhaustively. A systematic investigation on the lines of research herein indicated would be a valuable contribution to our knowledge.

'Crude Fibre.'—'Rohfaser.'—In connection with the lignocelluloses of cereals, the opportunity arises to discuss an artificial product with which agricultural chemists are familiar under the above description. In arriving at the nutritive value of food-stuffs it is necessary to discriminate between digestible and indigestible constituents. It has long been known that to the former belong chiefly the proteids, the water-soluble carbohydrates and fats; and to the latter, in general terms, the cellular tissue of vegetable food-stuffs. Between these two extreme groups lies the aggregate of compounds known as 'non-nitrogenous extractive matters.' It will be evident from discussions in this treatise (p. 86) that this complex admits of being resolved, by various processes of hydrolysis and oxidation, into carbohydrates of known constitution, or derivative products which determine the constitution of the groups from which they are formed. This aggregate is dissolved by treatment with weak hydrolytic agents, acid and alkaline, and the residue is the complex in question, known as crude fibre. A standard process for estimating this complex, which has been largely, in fact generally, used by agricultural chemists, is that known as the 'Weende method.' This consists in boiling the material to be analysed with dilute sulphuric acid (1·25 p.ct. H_2SO_4), and afterwards with dilute alkaline solution (1·25 p.ct. KOH), washing, drying, and weighing the residue. As the process of animal digestion may be briefly defined as an exhaustive series of hydrolyses under alternately acid and alkaline conditions, the method in question certainly gives a crude measure of the proportion of the material resisting the natural process of digestion. On the other hand, as an 'aggregate' method it is open to a good deal of objection; and, with the general advance of chemical and physiological methods of observation, the time has come for a revision of the subject, in order that the line separating 'digestible' from 'indigestible' matters may be defined more in accordance with directly ascertained facts.

In order to show in general terms the nature of the constituents 'digested,' i.e. dissolved by the artificial process, we give an abstract of a report upon 'Determinations of Crude Fibre and their Defects,' by C. Krauch and W. v. d. Becke, Landw. Vers.-Stat. 27, 5 (1882).

The residue from the treatments by the Weende method is generally assumed to be 'cellulose and woody fibre,' and, by inference, that these constituents resist the attack of the boiling acid and alkali. These authors determined the proportions dissolved from typical food-stuffs by the two treatments, together with the elementary composition of the aggregates, with the following results:

(a) Dissolved by the boiling dilute acid; (b) by the alkali; and (c) residue.

	(a)	(b)	(c)
Rye (grain)	52·12	26·48	21·40
Meadow hay	28·30	21·85	49·85
Clover hay	19·47	26·17	54·36

Elementary Composition of Aggregates

	(a)			(b)			Residue		
	C	H	O	C	H	O	C	H	O
Rye	47·6	6·03	46·36	55·12	7·68	37·23	55·11	7·58	37·03
Meadow hay	50·12	7·08	42·80	56·42	6·49	37·09	46·38	6·36	47·26
Clover hay	42·99	6·44	50·57	51·12	6·35	42·53	49·08	6·63	44·29

The above results are calculated with exclusion of the nitrogenous constituents (albuminoids) and ash. From the high C percentage of the constituents dissolved, it is evident that the lignocelluloses are attacked.

In more direct criticism of the assumed digestibility of the 'N-free extractive matters,' the authors investigated cereal 'meals.' The starch was estimated by the malt extract process, and the 'N-free extractives' by the Weende method, with the following results:

	(a)	(b)	(c)
Starch	62·48	42·02	26·61
N-free extractives	70·38	64·8	66·50

These specimens were selected in accordance with gradations in recognised feeding value from *a* to *c*, gradations corresponding approximately with the ascertained proportions of starch, but altogether at variance with the numbers for 'N-free extract.'

In further illustration of the same point the authors cite the following more complete analysis of meals (Brunner, Landw. Ztg. Westfal, 1877, p. 19).

	(a)	(b)	(c)
Proteids	15·56	15·89	17·35
Fat	2·53	2·74	5·63
N-free extractives	65·87	65·23	65·28
Crude fibre	8·33	9·17	6·84
Ash	7·71	6·97	4·90
Direct estimation of starch by malt method	27·93	30·4	53·63

It is again evident that the 'N-free extractives' are not a measure of the nutritive value; but, on the other hand, by a direct estimation of the starch, the method becomes more complete.

The authors then completed their investigation by taking as the basis of observation food-stuffs deprived of fats, by extraction with ether-alcohol, and starch, by digestion with water and malt extract at 50-60°. The residue, which they termed 'Grundsubstanz,' was then subjected to the Weende method of hydrolysis; and by determinations of elementary composition of the residues, the composition of the dissolved constituents was arrived at.

The specimens investigated were three grades of wheat-brans (pollards) and two specimens of rice meal. The materials operated on, viz. residues from the treatments above described, had the following composition:

	Brans			Rice meal	
	(a)	(b)	(c)	(d)	(e)
C	51·82	50·38	48·32	51·3	39·2
H	7·00	6·34	6·38	7·09	5·12
N	3·17	2·74	0·84	5·14	0·58
O	37·22	39·81	43·37	32·17	34·82
Ash	0·79	0·73	1·09	4·30	20·28

or, calculated to C,H,O compounds only—

	(a)	(b)	(c)	(d)	(e)
C	52·06	50·28	48·69	53·92	48·34
H	7·07	6·37	6·42	7·63	6·38
O	40·87	43·35	44·89	38·47	45·28

(*a*) The following were the results of the first treatment, boiling in 1·25 p.ct. H_2SO_4, in regard to the percentage and elementary composition of the N-free constituents dissolved:

		(a)	(b)	(c)	(d)	(e)
Proportion dissolved		43·07	49·62	52·15	20·72	8·84
Elementary composition	C	48·96	47·03	49·17	50·26	40·4
	H	6·34	5·46	6·48	7·46	8·78
	O	44·70	47·51	44·35	42·28	50·81

(*b*) And on subsequently boiling with dilute potash—

		(a)	(b)	(c)	(d)	(e)
Proportion dissolved		20·14	12·75	23·44	20·55	20·57
Elementary composition	C	57·17	57·63	50·83	57·62	53·44
	H	7·65	7·02	5·62	7·31	6 66
	O	35·18	35·35	43·55	35·07	39·60

It may be noted that the albuminoids are almost entirely removed by these treatments, as is evident from the determinations of N in the residual crude fibre, viz.:

$$N \quad 0·17 \quad 0·00 \quad 0·00 \quad 0·42 \quad 0·37$$

The conclusions drawn from these results are, that the constituents dissolved are of relatively high C percentage, that they consist in large proportion of the lignocelluloses of the raw material, and that therefore the residual crude fibre is a product of purely empirical, and in fact arbitrary value.

The subject is also exhaustively treated by A. Müntz, in a brochure entitled 'Recherches sur l'alimentation et sur la production du travail' (Ann. Agron. 1877-8, No. 2). This contains the results of a very elaborate inquiry into the muscular work of the horse in relation to the food consumed. The inquiry involved the determination of the nutritive value of typical fodders, as its necessary basis, and the author's conclusions in regard to the 'cellulose' constituents of these food-stuffs are noteworthy. After showing

that these all contain *celluloses* easily hydrolysed by dilute acids
to 'glucoses'—thus clearly anticipating the later work upon the
celluloses of Class C (p. 85)—the author makes the following
statement in regard to the cereal straws :—

'L'exemple le plus frappant de cet effet nous est offert par la
paille, dans laquelle le microscope ne nous fait découvrir aucune
trace d'amidon et dans laquelle pourtant on dose d'après les
méthodes ci-dessus indiquées 20 p.ct. d'amidon' : the method
consisting in hydrolysing with dilute sulphuric acid (2 p.ct H_2SO_4)
at 108°, and titrating, in alkaline solution, with Fehling's solution
as in glucose estimations. Celluloses thus easily hydrolysed may
be assumed also to be digestible by the animal organism, and to
have a value equal to that of starch. In the case of starchy fodders,
however, the author adds to his scheme a method for the exclusive
estimation of starch, and selects the process of diastatic conversion.
In determining *crude fibre* (described in the original as 'cellulose
brute') the method of alternate digestion (at 108°) with dilute acid
(2 p.ct. H_2SO_4) and alkali (5 p.ct. KOH) was adopted. The residues
from these treatments were found to have the composition, and
to be formed in the proportions, given in the annexed table :

—		Oats	Beans	Bran	Hay	Straw
Yield p.ct.		12·42	8·61	6·38	29·83	39·77
Elementary composition	C	45·55	45·00	49·93	48·05	47·62
	H	6·28	6·48	6·99	6·36	6·35
	O	48·17	48·52	43·08	45·59	46·03

the high C percentage being referred to the '*lignin*' group remaining in combination with the cellulose. The proportion of nitrogen
in these residues surviving the treatment is from 0·2 to 0·3 p.ct.—
which is neglected in the calculations. In criticising this process
the author clearly states that it is of purely statistical and approximate value, even as a measure of non-digestible constituents ; and
as a determination of cellulose, valueless, if not altogether misleading.

Having found in effect that from 20–25 p.ct. of the substance of
straws yields to acid hydrolysis in the same way as starch, and
that the constituent so attacked appeared to be of the nature of
cellulose, the following method is proposed for estimating the

cellulose proper (*cellulose réelle*) : The substance (straw) is digested with dilute hydrochloric acid (7 p.ct. HCl) for some hours, washed, and treated with strong ammonia until all matters soluble in this reagent are removed; it is then treated with cuprammonium solution. After exhaustive action of this solution, the cellulose is precipitated from the solution by the addition of acetic acid in slight excess, washed, dried, and weighed. The following results were obtained :

Process No. 1.

Cellulose estimated as described by solution in cuprammonium . 49·44

Process No. 2.

Cellulose dissolved in the process of estimating crude fibre . 21·99
Cellulose present in ' crude fibre ' (39·77 p.ct.) calculated from
  its composition[1] . . . . 26·54

Total . . . 48·53

The concordance of these numbers is perhaps misleading, since the lignocelluloses are attacked by cuprammonium. The errors of the two processes compensating one another, the author arrives at a determination of cellulose approximating to that of the now standard methods.

In dealing with the group of *substances indéterminées*, i.e. the residue of constituents not determined by the standard methods of analysis, the author arrives at conclusions similar to those contained in the paper above noted. By the statistical method he finds the composition of this complex in typical fodder plants to be as under:

	Oats	Maize	Beans	Bran	Hay	Straw
C . . .	46·8	45·5	44·1	45·6	51·3	47·7
H . . .	6·2	6·5	6·2	6·3	6·2	6·1
O . . .	47·0	40·0	49·7	48·1	42·4	46·2

numbers which indicate large variations in composition, and therefore in nutritive value.

The author's elaborate method of proximate analysis certainly effects a more complete resolution of this heterogeneous group ; and the complete scheme of investigation, devised to minimise the errors of the standard methods, is worthy of attention. It would

---

[1] *I.e.* as a mixture of cellulose and lignin.

take us too far from the purpose of this discussion to reproduce the scheme in full detail; it will be sufficiently grasped from the subjoined statement of the complete results of analysis of straw.

	Constituents estimated p.ct.	Proportion of elementary constituents			
		C	H	O	N
Hydrolysable cellulose (equivalent to starch)	21·99	9·77	1·36	10·86	—
Glucose	0·34	0·14	0·02	0·18	—
Fatty matters	1·26	0·97	0·15	0·14	—
Crude cellulose (crude fibre)	39 77	18·96	2·53	18·28	—
Pectic acid	0·89	0·37	0·04	0·48	—
Albuminoids	3·95	2·12	0·28	0·92	0·63
Sum of elementary constituents	—	32·33	4·38	30·86	0·63
Total, by elementary analysis of original straw	—	48·27	6·35	44·75	0·63
Differences = Elementary constituents of substances undetermined	31·80	15·94	1·97	13·89	0·00
Whence is deduced the percentage elementary composition of undetermined constituents	—	50·13 p.ct.	6·19 p.ct.	43·68 p.ct.	—
Total	100·00				

With the aid of the methods of more recent introduction (see p. 261), the group of 'undetermined constituents' of the older analytical schemes may be much more completely resolved; and these methods, added to those above outlined, afford a scheme of sufficient completeness for all the present requirements of agricultural or physiological research.

We have devoted some space to the consideration of these results, not only on account of the importance of the subject, but as an illustration of the statistical method of inquiry to which chemico-physiological investigations have been largely limited. By the later work on the chemistry of the more complex carbohydrates, the way is opened for direct investigation of the nutritive value of the group of constituents formerly aggregated as N-free extractives. The subject is of wide interest, involving questions of importance to the agriculturist and physiologist; but the methods by which it requires to be attacked are for the most part purely chemical, and such as are described in more or less detail in this treatise.

Having now dealt with the more prominent types of lignification in plants and tissues of 'annual' growth, it remains to deal with the lignocelluloses of perennial stems.

(3) **Woods and Woody Tissues.**—The stem of an exogenous perennial is a complex of structural elements of varied form and function. Of these we may distinguish three main groups: (1) vessels, (2) wood cells proper, and (3) medullary tissue. When compacted together to form the permanent woody tissue, these groups appear to be indistinguishable chemically; they all undergo 'lignification'; are characterised by the same reactions; and, although it has been stated that they are variously resistant to the action of destructive reagents, the variation has not been satisfactorily referred to any fundamental differences of composition. These points are well discussed by Sachsse in his Chemie u. Physiologie d. Farbstoffe, Kohlenhydrate u. Proteinsubstanzen (Leipzig, 1877), and we abstract a short *résumé* of his treatment of the subject.

To the action of the concentrated non-oxidising acids ($HCl.H_2SO_4$) the *vessels* are generally more resistant. This is especially the case, however, in the earlier stages of growth. Thus sections of fleshy roots (*Daucus Carota*), treated alternately with dilute potash and concentrated hydrochloric acid (2-3 days' digestion), are disintegrated by the treatment, the cellular tissue being entirely broken down; the vessels, however, survive, and are isolated free from the cellular matter in contact with which they were built up. But with older tissues, on the other hand, no such differentiation is observable: the three groups of structural elements are equally attacked by purely hydrolytic treatments, however, they may be varied, both as to reagents and conditions of action.

Similar results are noticed with oxidising agents. Thus chromic acid solution (20 p.ct.) attacks the parenchymatous tissue of young growths much more rapidly than the vessels, which, with a careful regulation of the treatment, may by its means be isolated more or less perfectly. In the woods the reagent appears to attack the vessels more rapidly than the wood cells and medullary tissue ; but any difference of action is not such as to permit of an isolation of the one or other group.

Schulze's reagent (HNO_3 and $KClO_3$) also attacks the several groups more or less uniformly, the differences noted being rather as between woods of different ages and species ; thus sap wood is more resistant than heart wood, and the 'soft' than the 'hard' woods.

In view of these results, the methods of classification adopted by Frémy will be seen to rest upon very insecure foundations. The basis of the classification is the resolution of the substance by successive treatment with HCl (dilute and conc.), $H_2SO_4.H_2O$, cuprammonium, alkaline hydrates, &c. Thus the following individuals have been isolated and defined as follows : cellulose, paracellulose (soluble in cuprammonium after treatment with acids), metacellulose (insoluble in cuprammonium) and vasculose (Frémy, Compt. Rend. 48, 862 ; Urbain, Ann. Agron. 9, 529). This classification has been severely criticised by Kabsch (Pringsheim, Jahrb. f. Wiss. Bot. 3, 357), and is entirely rejected by Sachsse (loc. cit.), Hugo Müller (Pflanzenfaser, p. 7), and other authorities ; and it is unnecessary to add anything to the criticisms of these writers.

There have been many attempts to resolve the woods into proximate constituents, but the authors have for the most part concluded, from their investigations, (1) that the fundamental tissue of the woods—i.e. the woods freed from adventitious constituents such as tannins, colouring matters, resins, &c.—is composed of substances which cannot be resolved by hydrolytic treatments into proximate components; and (2) that there is a striking uniformity in composition of the fundamental tissue of the woods, notwithstanding their structural complexity; and this uniformity embraces not merely the individuals of a species, but extends over the widest range of such products.

Mention should be made here of a *general property of the woods*, and lignocelluloses discovered and investigated by C. Wurster, viz. that of fixing atmospheric oxygen in the form of a *peroxide*, giving the reactions of the typical *hydrogen peroxide*. One of these is the oxidation of the methylated derivatives of paraphenylendiamine to red colouring matters—and this reaction is equally and generally characteristic of the woods. Wurster has, in fact, reduced the reaction to an approximate quantitative estimation of the proportion of 'mechanical wood pulp' in papers. The reagent in question is incorporated in definite proportion with a pure cellulose paper, which is used as a 'test paper'; the paper to be tested being moistened with water, and the test paper pressed into the moistened part. The depth of colour is compared with standard coloured papers constituting a scale, the oxidising effects producing the colour being also expressed in terms of normal iodine solution. The percentage of wood in the paper corresponds to the depth of colour produced on the test paper.

For an account of Wurster's researches, which have been extended to a number of organic products, and are of considerable physiological interest, see Berl. Ber. 1887, 20, 808 (Quantitative Bestimmung des Holzschliffes im Papier), and also pp. 256, 263, 1030, 2631, 2934. From these it appears that the reaction, in the case of the woods, is the expression of quinonic constitution of characteristic (hexene) groups.

EMPIRICAL COMPOSITION (ELEMENTARY ANALYSIS) OF WOODS.—The uniform composition of the woods has been for many years regarded as established on the basis of the analyses of Chevandier (Ann. Chim. Phys. [3], 10, 129), which are still retained in all text-books, and may therefore be reproduced here.

	Beech	Oak	Birch	Poplar	Willow
Carbon	49·89	50·64	50·61	50·31	51·75
Hydrogen	6·07	6·03	6·23	6·32	6·19
Oxygen	43·11	42·05	42·04	42·39	41·08
Nitrogen	0·93	1·28	1·12	0·98	0·98

A series of determinations has also been made by Gottlieb (J. Pr. Chem. [2], 28, 385). These include the elementary analyses of the woods, and the calorific equivalents in heat units per 1 grm. burned.

Inorganic		Organic			Calorific equivalent
Ash	Wood	C	H	N	Heat units per 1 grm.
0·37	Oak	50·16	6·03	—	4,620
0·57	Ash	49·18	6·27	—	4,711
0·50	Hornbeam	48·99	6·20	—	4,728
0·57	Beech	49·06	6·11	0·09	4,770
0·29	Birch	48·88	6·06	0·10	4,771
0·28	Fir	50·36	5·92	0·05	5,035
0·37	Pine	50·31	6·20	0·04	5,085
	Cellulose	44·4	6·2		4,146

These results have been extended by G. W. Hawes (Amer. J. Sci. [3], 7, 585) to the woody tissues of Acrogens—e.g. Lycopodium, Equisetum, Aspidium, &c.—and from his investigations he concludes that the same general relationship obtains for these as in the forest trees above given.

PROXIMATE ANALYSIS.—A large number of the woods are characterised by special constituents, more or less of the nature of excreta. To deal with these would lie outside the scope of the present treatment of the subject. We are strictly limited to the fundamental tissue of the woods, considered as lignocelluloses. Hugo Müller (Pflanzenfaser, 150) gives the results of analyses of a representative series, the most important numbers being the percentages of cellulose isolated by the BrAq method. His results are comprised in the following table:

Wood	Water	Cellulose	Aq extract	Resin	Non-cellulose
Birch	12·48	55·52	2·65	1·14	28·21
Beech	12·57	45·47	2·41	0·41	39·14
Box	12·90	48·14	2·63	0·63	35·70
Ebony	9·40	29·99	9·99	2·54	48·08
Oak	13·12	39·47	12·20	0·91	34·30
Alder	10·70	54·62	2·48	0·87	31·33
Lignum Vitæ	10·88	32·22	6·06	15·63	35·21
Lime	10·10	53·09	3·56	3·93	29·33
Chestnut	12·03	52·64	5·41	1·10	28·82
Fir	12·87	53·27	4·05	1·63	28·18
Mahogany	12·39	49·07	9·91	1·02	27·61
Poplar	12·10	62·77	2·88	1·37	20·88
Pine	13·87	56·99	1·26	0·97	26·91
Teak	11·05	43·12	3·93	3·74	38·16
Willow	11·66	55·72	2·65	1·23	28·74

An account of the applications of Frémy's method of proximate analysis to the woods will be found in J. Chem. Soc. 1884, 46, 860 (abstracted from Urbain, Ann. Agron. 9, 529-547).

Thus oak wood was purified by treatment with alcohol-ether (losing 4 p.ct.), and afterwards with water and weak alkaline solutions (losing an additional 10 p.ct.). The residue, or lignocellulose proper, is treated with cuprammonium to remove *cellulose*; the residue boiled with dil. HCl, and again digested with cuprammonium to remove *paracellulose*; the residue is *vasculose*. Analysed in this way, the lignocelluloses of oak and poplar give the following results:

	Cellulose	Paracellulose	Vasculose
Oak	27·05	42·90	30·05
Poplar	34·10	45·95	19·95

When this method has been employed in investigations, the results should be compared with the results of the methods adopted in this treatise. The reagents employed will be found to be not selective in their action according to the distinctive *constitutional* features of the component groups of the woods.

Schulze investigated the resolution of the woods into cellulose (insol.) and non-cellulose (sol. derivatives) by the process of digestion with nitric acid (dilute) and $KClO_3$ (p. 96). The cellulose was estimated and analysed; the following are representative results:

Wood	Elementary composition of isolated cellulose		Proximate composition of wood	
	C	H	Cellulose (estimated)	Non-cellulose (diff.)
Beech	47·71	6·05	48·41	51·59
Oak	44·51	6·00	45·87	54·13
Alder	43·96	5·95	47·97	52·03
Acacia	44·29	6·09	52·94	47·06
	44·54	6·01	58·11	41·89

A further consideration of these results by the statistical method, taking the empirical composition of the woods as the basis of comparison, led Schulze to adopt the formula $C_{19}H_{24}O_{10}$ (C = 55·3 p.ct.) as approximately representing the composition of the non-cellulose or lignone complex—a formula which is in very close agreement with that which we have adopted for the lignone of the typical lignocellulose.

N. Schuppe has also investigated the composition of woody tissue, upon similar lines and with similar results (Pharm-Journ. [3], 14, 52). He arrives at the formula $C_{19}H_{18}O_8$ for the lignone complex, and at the mixed expression

$$5C_6H_{10}O_5 \cdot C_{19}H_{18}O_8$$

as representing the average composition of woody tissue.

It is evident from these results that woody tissue is similarly constituted to the typical lignocellulose, the main difference being the higher percentage of carbon (higher proportion of lignone) and lower proportion of cellulose.

We have not as yet endeavoured to connect the process of lignification in any definite way with the products—viz. the lignocelluloses—but the moment has now arrived for briefly setting forth a general theory of the process. It is many years since Sachsse definitely propounded the view that the lignocelluloses are products of metabolism of cellulose.

Comparing cellulose, $C_{18}H_{30}O_{15}$, with lignone calculated to a C_{18} formula, viz. $C_{18}H_{24}O_{10}$, the latter could be formed from the former by dehydration ($-3H_2O$) and deoxidation ($-O_2$). Sachsse preferred to formulate the process hypothetically as under ('Farbstoffe,' &c. p. 146)—

$$\underset{\text{Cellulose}}{C_{24}H_{40}O_{20}} - C_6H_6O_5 - 5H_2O = \underset{\text{Lignone}}{C_{18}H_{24}O_{10}}$$

the unknown complex $C_6H_6O_5$ undergoing further change, either oxidation to CO_2, or condensation to aromatic products (tannins). This view is based entirely upon physiological evidence, the chemistry of the process being hypothesis, pure and simple. But we are now in a position to supply a more substantial and consistent chemical basis.

(1) The celluloses isolated from the lignocelluloses are all oxycelluloses—i.e. contain furfural-yielding groups. Oxidation of the celluloses upsets the molecular equilibrium; the oxyproducts are relatively unstable, and easily condensed to closed chain derivatives. The celluloses also contain, in certain cases, methoxyl groups.

(2) The non-cellulose contains, in addition to the closed hexene rings and methoxyl groups, a characteristic and condensed cellulose derivative. This group we can only diagnose by indirect means. But a careful review of the evidence leaves no doubt that the cellulose on the one hand, and keto R. hexene derivatives on the other, regarded as extreme members, are connected by an intermediate product or group of products, which can be transformed (*a*) into cellulosic derivatives; (*b*) into acid products of low molecular weight, chiefly acetic; (*c*) into closed ring compounds, amongst which furfural is prominent.

It is impossible to draw any line of separation between these main groups: the 'intermediate' constituent in some reactions (e.g. chlorination) remains united to the R. hexene groups ; in others passes into cellulose, and is isolated as such (regulated oxidation by CrO_3); in others it is broken down, and this destruction takes place at the expense of cellulose. These considerations are dealt with at greater length in a paper by the authors, 'Celluloses, Oxycelluloses, Lignocelluloses' (Berl. Ber. 1893, 2520).

(3) Regarding lignification as a process of continuous modification of cellulose, and the woods as representing the extreme limits of such a process, these should show an increase in lignone at the expense of cellulose; which is in fact the case. Lignocelluloses in the first year of growth contain 70-80 p.ct. cellulose ; the woods, on the other hand, 50-60 p.ct.

(4) The woods often contain aromatic products of definite and well-ascertained constitution. We have given some of the evidence for the view that the tannins in heart woods are direct products of transformation of the lignocelluloses of the tissue. We shall presently see that aromatic products are formed in large quantity in the process of destructive distillation of the woods, some of the most characteristic of these being pyrogallol derivatives. We have also seen that the origin of the hippuric acid of the herbivora has been traced to the lignocelluloses of the fodder plants. It is, therefore, generally established that the lignocelluloses are connected on the one hand with the celluloses, and on the other with the derivatives of benzene, through a series of intermediate products ; and in the present state of knowledge it is not difficult to account for the relationships which exist as genetic.

(5) The most convincing evidence, however, is that furnished by a general review of the numerical constants of the lignocelluloses. We have already alluded to the general uniformity in composition of the woods, and therefore for our immediate purpose we select a typical wood (beech) for

comparison, on the more essential points, with the typical 'annual' lignocellulose—viz. jute.

	Elementary composition		Proximate resolution		Quantitative reactions of non-cellulose		
	Carbon	Hydrogen	Cellulose	Non-cellulose	Methoxyl	Furfural	Cl combining
Jute	46·5	—	75	25	4·0	8·2	8·0
Beech	49·1	—	55	45	6·2	12·8	12·0

In regard to the number thus obtained for beech, i.e. after merely boiling in water, it is necessary to point out the cause of the difference from the number given on p. 195, viz. 8·0 obtained after exhaustion with alkali, and calculated on the product so exhausted. The alkaline treatment removes a complex made up of pentosan, acetic residue, and keto-hexene constituents, the removal of which gives a residue approximating in composition to the jute lignocellulose; whereas in the complex removed, the proportion of the R. hexene groups, which react with chlorine, is much higher; and hence the increased proportion of Cl combining.

It is unnecessary to enlarge upon the simple relationships of these numbers. Sachsse's theory of lignification, it must be remembered, was based upon purely physiological grounds, and could only be supported by the meagre evidence of such empirical determinations as were then available. Now that we bring to bear the results of quantitative determinations based upon specific constitutional features, and find these perfectly consistent with this theory, the time has arrived to press its consideration as a generalisation of wide import, concerning the constructive processes of the organic world. The theory briefly expressed, and in its more enlarged scope, is this: that the process of lignification consists in a series of progressive and intrinsic modifications of a cellulose or oxycellulose tissue, the products of modification remaining associated with the residues of the parent substance in a state of combination or of intimate mixture, the final products of metabolism (aromatic products, pentosans, &c.) being excreted and taking no further part in the organic processes of the tissue.

We pass now from general deductions as to the constitution of the woods to the consideration of the results of special investigations (*a*) of particular constituents or reactions of the woods as a class ; (*b*) of particular woods.

(*a*) With the advance of analytical methods, characteristic constituents of the woods may now be determined with precision. Estimations of furfural and methoxyl have been carried out in extended series, and the results are significant for their uniformity on the one hand, and for the indications of progressive variation with the progress of lignification on the other.

(1) ESTIMATIONS OF FURFURAL.—*Furfural-yielding constituents.*—In a recent publication of Professor Tollens (Zeitschr. Ver. f. d. Rübenzucker Industrie, 44, No. 460) he gives a *résumé* of results obtained by himself and others who have collaborated with him in the investigation of the furfural-yielding constituents of plant tissues. The following numbers have been selected as the mean results for the more important woods :

Wood	Yield of furfural p.ct.
Hard woods { Beech	12·6
Oak	10·7
Birch	13·7
Pine	5·0

De Chalmot has made a more extended series of determinations (Amer. Chem. Journ. 16, 224), the results of which are given in the table on p. 182.

There is therefore a striking uniformity amongst the woods

of the Dicotyledoneæ in regard to this constituent or group of constituents, and an equally striking divergence in the case of the woods of the Coniferæ. This differentiation of the Coniferæ will be dealt with in the section devoted specially to the group (p. 197).

Species	Family	
Liriodendron tulipifera	Magnoliaceæ	9·5
Magnolia acuminata	Magnoliaceæ	8·5
Prunus pennsylvanica	Rosaceæ	9·8
Cercis canadense	Leguminosæ	10·5
Acer dasycarpum	Sapindaceæ	11·0
Liquidambar styraciflua	Hamamelideæ	10·5
Ilex opaca	Ilicineæ	12·3
Ampelopsis quinquefolia	Vitaceæ	10·4
Cornus florida	Cornaceæ	10·8
Nyssa sylvatica	Cornaceæ	10·4
Fraxinus americana	Oleaceæ	8·7
Fraxinus platycarpa	Oleaceæ	8·7
Juglans cinerea	Juglandaceæ	9·6
Carya alba	Juglandaceæ	10·6
Salix spec.	Salicineæ	10·5
Betula spec.	Cupuliferæ	11·7
Fagus ferruginea	Cupuliferæ	10·5
Quercus Phellos	Cupuliferæ	10·8
Quercus alba	Cupuliferæ	10·2
Quercus rubra	Cupuliferæ	10·8
Quercus nigra	Cupuliferæ	10·7
Ulmus americana	Urticaceæ	8·7
Platanus occidentalis	Platanæ	10·8
Juniperus virginiana	Coniferæ	5·2
Pinus strobus	Coniferæ	3·7
Pinus mitis	Coniferæ	4·4
Tsuga canadensis	Coniferæ	3·0

De Chalmot has further investigated the question of the life-history of the woods in reference to this group of constituents, and has established the important conclusion that they preserve a more or less constant ratio to the entire woodsubstance. The determinations were made upon the wood of particular and successive rings in horizontal sections, the position of the rings in the section determining the age of the wood-substance. The following observations may be cited (*loc. cit.* p. 225):

Wood of Quercus nigra.

	Furfural
2–12 years old	10·6 p.ct.
69 ,,	10·5 ,,
109–110 ,,	10·3 ,,

Liriodendron tulipifera.

19–21 years old	9·8 p.ct.
59–60 ,,	9·4 ,,

Magnolia acuminata.

Year rings of this tree not distinctly discernible.

Younger than 10 years	8·8 p.ct.
Old heart wood	8·4 ,,

Platanus occidentalis.

4–10 years old	9·8 p.ct.
71–79 ,, (Heart wood)	9·7 ,,

Prunus pennsylvanica.

(1) 4–12 years old	10·8 p.ct.
(2) ,, ,,	10·4 ,,
(1) 60–70 ,,	9·8 ,,
(2) ,, ,,	9·6 ,,

In the above it will be observed there is a slight diminution in the furfural with age, but the maximum difference may be taken at 1 p.ct. on the wood, or approximately 10 p.ct. on the furfural-yielding constituents themselves.

In other specimens, on the other hand, a slight increase was observed, thus:

Juglans cinerea.

	Furfural
(1) 4–18 years old	9·1 p.ct.
(2) ,, ,,	8·9 ,,
(1) 60–65 ,,	9·6 ,,
(2) ,, ,,	9·5 ,,

Fraxinus americana.

2–7 years old	8·9 p.ct.
51–52 ,,	9·4 ,,

It is therefore established by these results that the furfural yielding constituents of the wood-substance undergo very little change with age. It is necessary to point out that De Chalmot uses the term 'pentosan' as identical with 'furfural-yielding compound,' but this requires some qualification. The formation of furfural is an empirical and an aggregate result, and, while specially characteristic of the pentoses, is also a property of certain oxidised derivatives of the hexoses, notably glycuronic acid. It is probable that the furfural may be formed immediately, in this case also from a C_5 derivative, a product of resolution of the glycuronic acid—a view which is supported by the observation that the acid when boiled with hydrochloric acid yields carbonic anhydride in quantity corresponding with the equation

$$C_6H_8O_6 = C_5H_4O_2 + CO_2 + 2H_2O,$$

viz. 26·5 p.ct. CO_2. The yield of furfural, on the other hand, is only 15·3 p.ct.; but this discrepancy may very well result from secondary condensations of the C_5 aldose (Mann and Tollens, *loc. cit.*), in consequence of which only the small proportions are decomposed in the second stage according to the equation.

The import of these qualifying considerations is, perhaps, rather physiological than chemical, showing that a number of minor changes may be taking place in the furfural-yielding groups without affecting their proportion to the lignocellulose as measured in terms of this end-product of their decomposition.

There is evidence that such changes do take place with age, resulting in the *formation of pentosans as such*. The dicotyledonous woods all contain the body known as *wood gum* (Holzgummi), which appears to consist, for the most part, of xylan. This substance is extracted by treatment of the ground wood (sawdust) with solutions of sodium hydrate (2-5 p.ct. Na_2O) in the cold; from the solution the 'gum' is precipitated on the addition of alcohol. The yield varies from 10-20 p.ct. of the weight of the wood. The gum is easily hydrolysed by boiling dilute acids with formation of xylose. Jute, on the other hand, gives only very small yields of this product, Tollens obtaining only 1·75 p.ct. by digesting the fibre with 5 p.ct. NaOH solution. This product also yields xylose on hydrolysis.

It is this difference of yield of the proximate product which requires to be emphasised, as it is altogether out of proportion to the relative yields of *furfural.* With the progress of lignification, in fact, there is probably a progressive formation of *pentosan* resulting from molecular changes within the particular group. These pentosans differ from the parent substance or complex in readily yielding to hydrolysis, and to this extent may be regarded as dissociated or split off from the fundamental tissue-substance—in other words, as excreta or end-products of metabolism. This view necessarily is involved in the wider question of the physiological significance of the furfural-yielding constituents of plants. De Chalmot has published a series of communications in elucidation of this question, under the titles 'Soluble Pentoses in Plants,' 'Pentosans in Plants,'&c. (Amer. Chem. Journ. 15, 16). One important result of these investigations is the conclusion that pentosans are not formed in any perceptible quantity by the assimilation process. This is equivalent to the statement that they must arise by secondary transformations of the hexoses before or after their elaboration into the permanent tissue of the plant.

Investigations of the germination process in relation to pentosans have given variable results: in some cases there is an increase in the total pentosan, in others a decrease; and in certain cases the pentosans of the seeds, e.g. of *Tropæolum majus*, appear to behave as 'reserve materials.' The authors have also investigated this question by studying the germination of barley. In the germination of barley there is not only an increase in the 'total pentosan,' but the early permanent tissue is found to contain a considerable proportion of these furfural-yielding constituents. Of these constituents, moreover, more than 80 p.ct. resist the process of alternate digestion in cold dilute acids and alkalis ; they are not therefore pentosans in the ordinary acceptation of the term. On the other hand, the pentosans proper are found in relatively large proportion in the later stages of growth of the cereal straws ; and, again, the evidence leads us to regard the pentosans as secondary products of metabolism, in contradistinction to primary products of assimilation. It is evident from this brief outline that the physiology of the pentosans—their origin, fate, and general significance—is still, in many directions, problematical. In regard, however, to the narrower problem of lignification, we may sum up the evidence as follows : The formation of furfural-yielding products invariably accompanies lignification. These products exist in the earlier stages of lignification in the cellulosic form, but with age (perennial stems) are gradually transformed into pentosans of relatively low molecular weight, and ceasing to occupy any organic relationship to the tissue. The proportion of these constituents is uniform (18–24 p.ct.) over a wide range of woods hitherto investigated, and varies, moreover, but little with the age of the wood ; the proportion is, however, much less in the woods of the Coniferæ (6–9 p.ct.), which therefore represent lignification of another chemical type (see p. 197). So far no relation has been traced between the percentage of 'pentosan' and the physical properties of the woods.

Before passing from this section of the subject we must describe somewhat more in detail the characteristic product, wood gum, already briefly noticed.

Compound Celluloses

Wood gum was first isolated and investigated by T. Thomsen (J. Pr. Chem. [2], 19, 146), and Poumarède and Figuier (Annalen, 64, 388). It is obtained from the woods of the ash, elm, oak, beech, willow, cherry, &c., by digestion with solutions of the alkaline hydrates as already described. The following particulars of later investigations by Wheeler and Tollens (Landw. Vers.-Stat. 39, 437) are noteworthy. After extraction from beech wood, and precipitation by alcohol, the product is purified by digestion with alcohol and hydrochloric acid, and washing first with alcohol and then ether. It is obtained thus as a white powder. In alkaline solution it exhibits strong lævo-rotation $(a)D = -69^\cdot6°$. The yield is approximately 15 p.ct. of the wood. Hydrolysed with boiling acids it gives a large yield of crystallisable xylose. Cherry wood under the same treatment yields 12–13 p.ct. of the product, also yielding xylose as the chief product of acid hydrolysis. It is noteworthy, on the other hand, that 'cherry gum,' the well-known exudation from the tree, yields the isomeric pentaglucose arabinose as the chief product of hydrolysis.

The cereal straws also yield, under similar treatment, 14-17 p.ct. of the product, but retaining a large proportion of the inorganic constituents of the straw, chiefly silica. This product shows a stronger rotation, viz. $(a)_D = -84^\cdot1°$. With the aid of heat in the alkaline digestion, a much larger yield of the product (26 p.ct.) is obtained.

Wood gum is insoluble in cold water, but slowly dissolves on boiling with water; on cooling, the solution is strongly opalescent; but on the addition of alkali in small proportion, a perfectly clear solution is obtained. The compound is insoluble in aqueous ammonia, and, in the process of isolating it, the raw materials are therefore usually subjected to a preliminary digestion with dilute ammonia, which removes colouring matters, &c.

Numerous (elementary) analyses of wood gum have given

numbers corresponding approximately with the empirical formula $C_6H_{10}O_5$, which is confirmed by more recent results (Tollens). The only value of these results, however, is to establish a normal 'carbohydrate' formula. The more important problem of its *constitution* has been elucidated, as already noted, by its yielding the pentaglucose xylose as the main product of proximate hydrolysis (acid), and furfural as the ultimate product (HCl). The most recent analyses of Tollens gave the following results :

		Furfural	Xylose
Wood gum	Specimen i	38·78	74·26
	,, ii	46·90	89·82
	,, iii	48·08	92·02
	,, iv	33·30	63·73

These specimens were from beech wood, variously prepared : No. i by the process already described ; Nos. ii and iii by extraction with alkaline solution, after boiling the raw material with dilute sulphuric acid ; No. iv by extraction with boiling milk of lime. It is evident from these results that wood gum is a pentosan—the amorphous anhydro-aggregate of xylose or xylan—mixed or combined with variable proportions of a carbohydrate of similar empirical composition, probably a cellulose derivative. It also generally contains methoxyl (2·6 p.ct. $O.CH_3$). These observations further confirm the view that the pentosans are derived from hexose groups, and represent the final terms of a series of transformations of which 'wood gum' as directly obtained may be taken as representing the intermediate terms.

METHOXYL DETERMINATIONS.—The $O.CH_3$ group, as a *chemical constant of lignification*, has been brought into prominence by the investigations of Benedikt and Bamberger ; their most important communication on the subject appearing under the title, ' Ueber eine quantitative Reaction des Lignins' (Monatsh. 11, 260-267). Employing the perfected method of

Compound Celluloses

Zeisel, these observers have made an elaborate series of estimations, the results being expressed as percentages of methyl (CH_3) calculated on the dry substance. They are as under:

A. Woods.

				CH, p ct.
Maple	Stem		*Acer Pseudo-platanus*, L.	3·06
,,	,, extracted [1]		,, ,,	3·05
,,	,, shavings		,, ,,	3·06
Acacia	Branch		*Robinia Pseud-Acacia*, L.	2·37
,,	Extracted		,, ,,	2·45
Birch	3 years old		*Betula alba*	2·57
Pear	Stem		*Pyrus communis*, L.	3·21
Oak	,,		*Quercus pedunculatus*	2·86
,,	,,		,, ,,	2·63
Alder			*Alnus glutinosa*	2·89
Ash	Stem		*Fraxinus excelsior*, L.	2·71
,,	Shavings from stem		,, ,,	2·69
,,	Stem shavings extracted		,, ,,	2·66
,,	Shavings from branches		,, ,,	3·02
,,	Shavings from branches extracted		,, ,,	2·91
Fir	Stem		*Abies excelsa*	2·15
,,	,,		,, ,,	2·25
,,	,,		,, ,,	2·39
,,	,, (central zone)		,, ,,	2·59
,,	,, (sap wood)		,, ,,	2·32
,,			*Abies pectinata*, DC.	2·45
Pine			*Pinus sylvestris*, L.	2·25
,,	Stem		*Pinus laricis*	2·05
,,	,,		,, ,,	2·12
Cherry	,,		*Prunus Avium*, L.	2·38
Larch	,,		*Larix europæa*, DC.	1·99
,,	,,		,, ,,	2·68
Lime	,,		*Tilia parvifolia*	2·56
Mahogany	,,		*Swietenia Mahagoni*, L.	2·66
Walnut	,,		*Juglans regia*, L.	2·27
,,	Shavings from stem		,, ,,	2·69
Poplar	Stem		*Populus alba*	2·59

[1] 'Extracted' signifies previously exhausted with water, alcohol, and ether. Otherwise the specimens were analysed without previous preparation.

								CH,p.ct
Beech	.	Stem	.	.	.	.	*Fagus sylvatica* .	. 3·02
,,	.	,,	.	.	.	.	,, ,, .	. 2·62
,,	.	,, shavings	.	.	.	,, ,, .	. 2·70
Elm	.	,,	.	.	.	.	*Ulmus campestris* .	. 2·92
,, .	.	,, shavings extracted .	.	,, ,, .	. 2·75
Willow	.	.	.	.	.	.	*Salix alba* .	. 2·31

B. Fibrous Products.—Natural and prepared.

Jute (*Lignocellulose*)	.	.	.	.	.	.	.	. 1·87
Swedish filter paper	.	.	.	.	.	.	.	. 0·0
Cotton	.	.	.	.	.	.	.	. 0·0
Flax, unbleached	.	.	.	.	*Linum usitatissimum*	. 0·0
Hemp	,,	.	.	.	.	*Cannabis sativa*	.	. 0·29
China grass,,	.	.	.	.	*Böhmeria nivea*	.	. 0·07
Sulphite (*Cellulose*)	.	.	.	*Pinus sylvatica*	.	. 0·34

C. Miscellaneous.

Cork .	.	.	.	.	.	*Quercus suber*	.	. 2·40
,, .	.	.	.	.	.	,, ,,	.	. 2·47
Nutshells .	.	.	.	.	*Juglans regia*	.	. 3·74
Lignite (Wolfsberg)	.	.	.	.	.	.	. 2·44
Brown coal	.	.	.	.	.	.	.	. 0·27

From these determinations it is evident that the formation of methoxyl groups is an essential feature of lignification, and, moreover, that the formation takes place with remarkable uniformity over a wide range of woody tissues. This uniformity is, indeed, such that Benedikt and Bamberger proposed to adopt the 'methoxyl number' as the quantitative measure of any wood lignocellulose present in an unknown fibrous mixture, e.g. for determining the proportion of 'mechanical wood pulp' in papers. From the above table it would be easy to calculate the degree of approximation (probable error) to be attained, and we may be satisfied to note that the approximation is sufficiently close to make such determinations distinctly valuable for the purpose in question. These authors were also enabled to draw from their results certain conclusions of physiological significance, viz.: (1) there is in the woods a slight progressive increase of methoxyl with age ; (2) there is a higher proportion of methoxyl in the wood of the branches as compared with the main stem ; (3) the proportion of methoxyl is unaffected by 'extracting' the wood, i.e. it is a characteristic constituent of the wood-substance (lignocellulose) itself.

THE ACETIC RESIDUE.—Acetic acid is produced in a number of the decompositions of the lignocelluloses (*ante*, p. 160). It is obtained more readily, and in larger proportion, from the (dicotyledonous) woods than from jute (*et similia*). The following reactions producing acetic acid may be cited :

(1) *Alkaline hydrolysis.*—The solutions obtained by treating beech wood with dilute aqueous alkalis contain acetic acid (acetate of soda), which is separated by distillation after acidification. The proportion is large, amounting to 7-8 p.ct. on the wood.

(2) *Acid hydrolysis.*—Acetic acid is formed on digesting the woods with dilute sulphuric acid at 60-100°. Larger yields are obtained by dissolving the wood-substance in concentrated sulphuric acid in the cold, diluting and distilling.

(3) *Oxidising processes.*—(*a*) *Acid.*—The wood, in fine shavings, is covered with normal sulphuric acid, and oxidised at ordinary temperatures, with its own weight of chromic acid (CrO_3) added in successive quantities. The solution on distillation yields acetic acid, equal to 5-6 p.ct. of the weight of the wood (dicotyledonous). The following are the results of actual determinations :

Beech	Sycamore	Birch
5·0 p.ct.	5·2 p.ct.	6·0 p.ct.

Oxidised with dilute nitric acid (10 p ct. HNO_3) at 60-100° (*ant.*, p. 146), very much larger quantities of acetic acid are obtained, viz. from 10-15 p.ct. of the weight of the wood.

(*b*) *Alkaline.*—The maximum yields are obtained in the drastic decomposition, determined by heating with the alkaline hydrates at 200–300°. The quantity obtained in this way is from 30–40 p.ct. of the weight of the wood, together with a considerable quantity of oxalic acid.

(4) *Destructive distillation* of the woods (see p. 204) also determines the formation of acetic acid. The following estimations of comparative yields are given by W. Rudnew (Dingl. J. 264, 88 and 128), the woods being 'distilled' in glass vessels at 150–300°.

Linden	10·24	Oak	7·9
Birch	9·5	Pine	5·6
Aspen	8·06	Fir	5·2

Wood celluloses (birch and pine), isolated by the Schulze process, gave under similar conditions the following yields :

Birch cellulose	6·2
Pine ,,	5·0

From these results it is evident that the $CO.CH_2$ grouping is a characteristic constitutional feature of the lignocelluloses. It also occurs in derivative forms amongst the products of decomposition of the lignocelluloses by 'natural processes. Thus, e.g., in hippuric acid, benzoyl-amido-acetic acid (p. 151), and in phloroglucol, regarded as $3CO.CH_2$, which occurs in the plant world in a number of derivative forms, and is obtained from several of the natural tannins as a product of fusion with alkaline hydrates. We are not yet in a position, however, to localise the $CO.CH_2$ groups in the complex lignocellulose molecule, and we cannot go beyond a summing-up of the evidence in general terms.

(1) Acetic acid is a product of simple hydrolysis, both acid and alkaline, of the lignocelluloses, the proportion being from 3–6 p.ct. of the parent substance.

The formation of an *acetic* residue is thus a characteristic feature of lignification. If derived from a hexose group (cellulose), it should be formed correlatively with the furfural-yielding compounds; and the quantitative relations of the two certainly confirm this view. Thus the hypothetical decomposition may be formulated as under:

$$2C_6H_{12}O_6 = 2C_5H_{10}O_5 + C_2H_4O_2,$$
$$\phantom{2C_6H_{12}O_6 = }2\times15060$$

and the pentosans of wood represent in effect a percentage approximately five times that of the acetic acid obtainable by simple hydrolysis.

In the jute fibre also, the smaller proportion of the furfural-yielding constituents is associated with a similar smaller proportion of the acetic residue. The formation of both therefore increases, *pari passu*, with age, which is in accordance with the view of a common origin.

(2) The celluloses, and the 'carbohydrates' generally, are susceptible of the 'acetic condensation.' The normal celluloses, however, require the application of drastic treatments, e.g. fusion with alkaline hydrates or warming with concentrated sulphuric acid, both of which treatments are of an oxidising character. The *oxycelluloses*, on the other hand—notably the straw celluloses—give a considerable yield of acetic acid (together with furfural) on long boiling with 10 p.ct. sulphuric acid. The maximum yield is obtained by dissolving the oxycellulose in the concentrated acid in the cold, diluting and distilling. In this way the authors have obtained a yield of 9-10 p.ct. of the acid, calculated on the oxycellulose.

These observations confirm the view that lignification is a process of transformation taking place in oxidised celluloses, or oxycelluloses, and following as a secondary result of the disturbance of equilibrium set up by the oxidation.

(3) In addition to acetic residues—converted by hydrolysis into acetic acid—there appears to be a $CO.CH_2$ nucleus, a dehydracetic residue, which is the source of the increased

yields of acetic acid under the action of dilute nitric acid. Of this constituent of the non-cellulose groups we have some indirect knowledge. Thus, in the case of jute, we have given to the entire lignone complex the statistical formula $C_{19}H_{22}O_9$. A portion of this, reacting with chlorine to form mairogallol, may be approximately formulated as $C_{18}H_{18}O_9 = 3[C_6H_6O_3]$.

If we therefore resolve the complex into

$$C_6H_6O_3 \quad 2OCH_3 \quad C_{11}H_{10}O_4,$$
Keto R. hexene group   Methoxyl

we are left with the highly condensed group $C_{11}H_{10}O_4$, containing the furfural-yielding constituents, and also yielding acetic acid as described.

The constitution of this complex must be considerably removed from that of the ordinary carbohydrates. Whether hexoses or pentoses are represented, either must be in the form of a polyanhydride; and the acetic residues are also probably of the dehydrated or $CO.CH_2$ form.

The further investigation of this problem is the work of the immediate future, and it is with the view of setting forth some of the probabilities involved that the discussion has been pushed somewhat beyond the limits of ascertained fact.

THE CHLORINATION REACTION.—The reactions of the wood lignocelluloses with chlorine have not been systematically investigated. It must be remembered that a wood tissue is a complex structure, and although it will have become evident that there is a remarkable uniformity in chemical composition, still a mixture is always less attractive as a basis of investigation than a homogeneous substance such as the jute fibre. It has, however, been sufficiently established by research that the reaction of the dicotyledonous woods with chlorine is identical in general features with that of the typical lignocellulose—i.e. a yellow-coloured quinone chloride is formed, giving the same brilliant colour reaction with sodium sulphite; and on treatment with alkali there is a complete resolution into cellulose (insoluble) and soluble derivatives of the lignone complex.

The coniferous woods, on the other hand, react somewhat differently, the chief distinctions being that the wood-substance is changed in colour to an orange red, and the product does not give any marked colour reaction with sodium sulphite. In both cases the percentage of chlorine combining with the lignocellulose is the same as with jute, viz. 8·0 p.ct.

Comparative experiments upon four typical woods gave the following statistics of reaction with chlorine. The results are given in terms of the lignocellulose proper, i.e. the residue from exhaustion with the alkaline solution (1 p.ct. NaOH).

		Pine	Beech	Sycamore	Birch
(a)	Residue from alk. treatment, or lignocellulose : p.ct. on wood	89	82	84	87·5
(b)	Cl. combining : p.ct. on (a)	7·5	7·5	9·0	7·0
	Acidity after chlorination calc. as HCl	23·5	19·5	21·0	15·0
(c)	Cellulose : p.ct. on (a)	72·0	65·0	70·0	72·5

These must be regarded as preliminary results, but they serve to confirm the view we have taken of the general and close similarity of the woods to the typical jute lignocellulose. It has not been determined whether the whole of the 'acidity' developed in the above chlorinations is due to HCl, or to acid products (e.g. acetic acid) split off from the lignocellulose.

The chlorinated derivatives have not been closely investigated. The authors have isolated one of these products obtained from a Spanish mahogany, the chlorination being preceded by the usual treatment with boiling dilute alkali (1 p.ct. NaOH). This product was found to contain 30·4 p.ct. Cl.

In regard to investigations involving the chlorination of these lignocelluloses, two points must be borne in mind : (1) As regards preparation of the material. To ensure a complete reaction the wood must be reduced to the finest possible shavings. (2) In regard to the preliminary treatment with boiling alkali. The woods are not attacked as a whole as with the jute fibre, the furfural-yielding constituents (pentosans) yielding much more readily than the fundamental tissue or lignocellulose proper. In systematic investigations following the lines laid down in the case of the jute fibre, the latter should be taken as the basis of observation, and not the *entire wood* substance.

The reactions which we have so far discussed are, in the main, reactions of decomposition. Synthetical reactions of the wood lignocelluloses have been still less investigated. Here, again, there is little to attract the chemist in the present state of our knowledge, owing to the necessary complexity of the reactions involved. From such reactions as have been studied, if only in a general and superficial way, it appears that the proportion of reactive OH groups is still less in these lignocelluloses than in those of which jute is the type. Thus, to select the reaction of nitration: The maximum yield of nitrate is considerably lower in the woods than in jute; moreover, the reaction is complicated by a destructive oxidation which supervenes at a very early stage of exposure to the action of the mixed acids. The following series of determinations of yield in the case of mahogany wood illustrate this point. In (*a*) the wood was used in its raw state; in (*b*) it was previously purified by boiling in dilute alkaline solution.

Nitrating acid: equal volumes of H_2SO_4 and HNO_3 (1·43 sp.gr.) in excess.

Duration of exposure to acid	Yield of nitrated wood	
	(*a*)	(*b*)
Mins.	P.ct.	P.ct.
1	106·6	115·6
2	118·4	121·0
3	126·5	127·2
4	112·7	125·3
5	108·8	123·1

After three minutes' exposure, therefore, in both cases oxidation supervened, accompanied by conversion into soluble products; this destructive oxidation being much more marked in the case of the raw wood substance. Jute, under similar conditions of treatment, would have given a maximum of 145 p.ct., and the nitrate is much more resistant to the continued action of the acid mixture.

These results are, of course, of slight value only; but they serve to give emphasis to the general conclusion that lignification is a process of condensation and etherification of OH groups, accompanied, and in part conditioned, by condensation in regard to carbon configuration. Similarly, also, the woods show considerably more resistance to the actions of solvents of cellulose than jute lignocelluloses; notably to the thiocarbonate reaction, to which they yield only in very slight degree and after prolonged exposure.

From this general view of the reactions of the woods considered as a class of the lignocelluloses, we proceed to consider special investigations of particular woods.

Woods of the Coniferæ.—These woods are of very great industrial importance, not merely for their uses as such, but as the raw material for the preparation of the 'sulphite wood pulp,' now produced on an enormous scale in connection with the paper industry. The ultimate fibres of these woods are of greater length than those of the dicotyledonous woods; in addition there are well-marked features of distinction in chemical composition from the latter, which have already been noted.

The chemistry of these woods was investigated some years ago by Erdmann (Annalen, Suppl. 5, 223). The wood of *Pinus abies*— purified from adventitious constituents by boiling in acetic acid, and subsequent exhaustion with water, alcohol and ether—gave, on ultimate analysis, constant numbers, viz.

C, 48·4 ; H, 6·3 p.ct. From these results, together with general observations on the chemical behaviour of the substance, it was concluded that it is a homogeneous compound, having the empirical formula $C_{30}H_{46}O_{21}$. Erdmann further concluded that this compound is resolved by hydrolysis with dilute acids into glucose (soluble) and a residue $C_{26}H_{26}O_{11}$, which he terms *lignose*, and the original compound therefore *glycolignose*. Lignose was further found to yield, on fusion with potassium hydrate, pyrocatechol and protocatechuic acid.

These results, or rather the interpretation of them, is inconsistent in many respects with the results of subsequent investigations. The experimental facts, however, remain ; and the researches are worthy of notice, as one of the earliest attempts to elucidate the constitution of the lignocelluloses as a definite chemical problem.

The 'sulphite pulp' process would appear to offer a much more promising field of investigation, since it not only determines a satisfactorily sharp separation of cellulose (pulp) from non-cellulose (soluble sulphonated derivatives), but with the minimum of chemical modification of either group. Notwithstanding these specific advantages of the process, considered as a method of proximate analysis, and numerous investigations of the soluble by-products ('sulphite liquor'), the constitution of the latter, and therefore of the original lignocellulose, still has to be expressed in very general terms. The most important contribution to the subject is that of Lindsey and Tollens (Annalen, 267, 341), of which the following is a brief account. The solution used in these researches was that resulting from the 'Mitscherlich process,' which consists in a prolonged digestion of the wood—after subjection to a preliminary mechanical disintegration—with a solution of calcium bisulphite. The solution is usually prepared to contain CaO 1·35 p.ct., SO_2

4.4 p.ct., and is used in the proportion of 5-7 parts to 1 part of wood. In the digestion, the temperature is gradually raised to 160° C.

The particular specimen of 'waste liquor' (1·055 sp.gr.) used in the above researches contained 9·5 p.ct. of 'total solids in solution' (dried at 100°), of which 0·58 p.ct. was CaO. The solution has a pale brownish-yellow colour, and reduces Fehling's solution strongly.

A systematic examination for the presence of carbohydrates of low molecular weight and known constitution gave for the most part negative results as follows :

(*a*) On boiling with HCl (16 p.ct. on the solution) after evaporation to a suitable volume, *traces only of levulinic acid* were obtained ; showing the general absence of such carbohydrates. (Annalen, 243, 333 ; Berl. Ber. 22, 370.)

(*b*) On oxidation with nitric acid *no saccharic acid* was formed, showing the absence of dextrose or dextrose-yielding compounds. (Annalen, 249, 222.)

(*c*) On oxidation with nitric acid, *traces of mucic acid* were obtained, showing the presence of galactose (or galactan) in small proportion. (Annalen, 232, 186, 205.)

(*d*) The solution was acidified with sulphuric acid, boiled some time, neutralised ($CaCO_3$), filtered, evaporated to a syrup, and boiled with strong alcohol. The clear solution was poured off, the alcohol evaporated, and the resulting syrup mixed with phenylhydrazine acetate. An insoluble hydrazone was obtained, which proved to be *mannose hydrazone*. An approximate estimate of the quantity showed 0·5-0·8 p.ct. on the solution, i.e. about 6-7 p.ct. of the 'organic solids.'

(*e*) On 'distillation' with hydrochloric acid, furfural was formed in some quantity. After precipitation of the bulk of the organic substances in solution with lead oxide, a solution was obtained which gave the brilliant colour reactions of the pentoses, and xylose was identified in the solution by precipitation with phenylhydrazine.

(*f*) Experiments were also made with the view of determining alcoholic fermentation (yeast) of the dissolved compounds. Small quantities of alcohol were obtained, but the maximum yield corresponded to 1·2 p.ct. only of carbohydrate.

The major proportion of the dissolved organic substances was found to be a gummy body with the usual ill-defined physical properties of the class of organic colloids. On the other hand, this body behaved, in many respects, as a homogeneous complex; and although it was found impossible to resolve it into proximate constituents of definite characteristics, it yielded a number of synthetical derivatives, from the analysis of which, compared with that of the gum itself, the conclusion resulted as to the homogeneity of the complex. The complex was obtained in various forms as follows:

(1) As a precipitate on adding hydrochloric acid to the original liquor.

(2) As a lead compound by precipitation of the wood liquor with lead acetate.

(3) The lead compounds were decomposed by treatment with sulphuric acid, and the solution treated with alcohol. A portion of the gum was precipitated as a flocculent mass, and a second fraction was obtained on evaporating the filtrate.

(4) A brominated derivative was obtained by treating the original wood liquor with bromine.

The following empirical formulæ represent the results of ultimate analyses of these products, together with methoxyl determinations.

(1) From analyses of HCl precipitate:

$$C_{24}H_{24}(CH_3)_2SO_{12}.$$

(2) Calculated from analyses of PbO compounds:

$$C_{24}H_{24}(CH_3)_2SO_{12}.$$

(3) From direct analyses of gums obtained from PbO precipitates:
- (a) Precipitated by alcohol: $C_{24}H_{24}(CH_3)_2SO_{12}$.
- (b) Soluble in aqueous alcohol: $C_{24}H_{24}(CH_3)_2SO_{12}$.

(4) Brominated derivative:

$$C_{24}H_{22}(CH_3)_2Br_4.SO_{11}.$$

The S is present in this complex as a sulphonic residue (SO_3H); the parent molecule, i.e. the non-cellulose or lignone complex of pine wood, may be regarded as approximately $C_{24}H_{24}(CH_3)_2O_{10}$. Certain definite conclusions may be drawn from this empirical study of its derivatives.

(1) It is evident that it represents a highly condensed molecule. Taking the 'carbohydrate' formula to which it most nearly approximates, viz. $C_{24}H_{24}O_{12}$, this represents

$$4C_6H_{12}O_6 - 12H_2O.$$

In addition to condensation expressed by dehydration, CH=CH groupings are also represented, as appears from the bromination of the product. The authors not having prepared the corresponding chlorinated derivative, we are not able to compare the grouping of the hexene rings of this complex with those of jute, especially as the reactions of the chlorinated wood are distinct from those of jute. There is, however, an unmistakably close general similarity.

(2) It is evident that the condensation is of a type which resists hydrolytic treatment of a very energetic character; and that the constituent groups of the lignocellulose are united together by stronger bonds of synthesis than O-linking.

(3) With regard to the mechanism of the reaction in the original resolution of the lignocellulose, it is of a complex character; and the synthetic equilibrium of the products in the resulting solution is, no doubt, different from that represented by the parent complex. We may very well assume that the reaction involves the following factors: (*a*) the hydrolytic action of the sulphurous acid; (*b*) the formation of aldehyde bisulphite compounds; (*c*) the probable sulphonation of side chains of the general form x.CH : CH.COH, as in the well-known reaction of cinnamic aldehyde with sodium bisulphite; (*d*) the saturation of acid OH groups by CaO.

These researches are, it will be seen, an important preliminary elucidation of the problem of the composition of this interesting industrial product, and afford general conclusions as to the constitution of the non-cellulose constituents of the lignocelluloses, which entirely confirm the deductions given in preceding sections of this treatise.

In regard to the pulp or insoluble product of the original reaction, which is, as already stated, an industrial product of the greatest importance, it represents the cellulose of the wood together with residues of the non-cellulose in small proportion, not removed by the treatment. The presence of the latter is marked by the colour of the product, which is usually a greyish-pink. A large quantity of the pulp is used in this crude or unbleached condition; but for white papers a preliminary treatment with bleaching powder is practised, the proportion required being from 15-25 p.ct. of the weight of the pulp. The process is attended by a loss of weight of from 8-12 p.ct., owing to conversion of the more oxidisable constituents of the pulp into soluble derivatives.

It is to be noted that the yield of bleached cellulose by this process is, as in many other cases, considerably inferior to that obtained by the process of chlorination, &c. By the latter the authors obtain 60-65 p.ct. of cellulose from the coniferous woods, whereas the 'sulphite process' yields about 50 p.ct. This is another instance of the variable character of the 'cellulose constants' of fibrous products; the cellulose being a product of resolution or decomposition, and varying both in character and proportion with the conditions of the treatment by which isolated.

With the exception of the woods of the Coniferæ, none of the woods have been submitted to exhaustive investigation so far as regards the fundamental tissue or lignocellulose proper. It appears, in fact, that such investigations have only been rendered possible by the general advances of the science during the last few years, more especially in the province of the carbohydrates. This work upon the carbohydrates of lower molecular weight, together with the preliminary work upon the general features of lignification recorded in this treatise, opens out a very wide field for future work in the direction of reducing the phenomena of elaboration and metabolism in the plant to exact molecular expression. In regard to such investigations we may point out here that of those types of lignification which have been so far studied, four may be selected as showing well-marked features of differentiation, viz. :

'Annual' products	'Perennial' products
Cereal straws; Jute bast	Dicotyledonous woods; Coniferous woods

each and all of which call for extended investigation, i.e. individually as presenting a problem of chemical constitution, and comparatively with the view to connect the variations in composition with variations in the physiological factors of their origin and growth.

There are, of course, a number of woods characterised by the secretion or excretion of particular products, such as more particularly the dye woods, logwood, brazil wood, sapan, &c., &c. These characteristic products are well-defined, mostly crystallisable compounds, the constitution of which is determined entirely without reference to the physiological problem of their origin or their relationship, genetic or otherwise, to the tissues in which they are stored up.

The purpose of this treatise is, however, strictly limited to the chemistry of fundamental tissue; outside this lies the indefinitely wide territory of plant secretions into which we make no attempt to enter.

We have now, in concluding our account of the lignocelluloses, to deal briefly with certain industrial processes which throw further light on the chemistry of the lignocelluloses.

(1) DESTRUCTIVE DISTILLATION.—The products of the destructive distillation of the woods are extremely numerous and of varied constitution, comprising, in fact, representatives of all the more important groups of C,H, and C,H,O compounds. The formation of these products depends upon various factors: (*a*) the composition of the wood itself, and (*b*) the conditions of distillation.

Ramsay and Chorley have made careful comparative investigations of typical dicotyledonous woods—oak, beech, and alder—and their results afford a general idea of the influence of these factors. The tables on the opposite page may be cited in illustration (J. Soc. Chem. Ind. 1892).

These results, as regards the solid residue (charcoal) and gaseous products and their relation to the conditions of distillation, are very complete and require no further discussion. The increase of gas at the higher temperature of distillation is formed at the expense of the charcoal, and CO at the expense of CO_2.

In addition to these observations on the products, the authors also found that the distillations were marked, as in the

	Oak	Beech	Alder
Weight of wood taken in grms.	167	180	134
P.ct. of charcoal	24.55	26.66	25.37
P.ct. of distillate	58.69	59.33	59.70
P.ct. of CO_2 absorbed by KOH	9.58	9.23	9.70
Difference to make up 100 p.ct.	7.18	4.78	5.23
Volume of gas after absorbing CO_2	7,000 c.c.	7,200 c.c.	6,000 c.c.
P.ct. composition of this gas — CO	70.77	73.14	73.47
O	—	1.02	1.52
Olefines	1.11	1.49	1.59
CH_4	14.90	18.71	20.11
N by difference	13.32	5.64	4.31
P.ct. of pitch from distillate on the wood	9.58	11.11	15.67
P.ct. of acetic acid	6.13	6.54	5.90
P.ct. of methyl alcohol	1.36	6.08	11.17

Maximum temperature in each case about 500°.

	Oak	Beech	Alder
Weight of wood taken in grms.	181	187	150
P.ct. of charcoal	33.7	34.22	34.66
P.ct. of distillate	56.35	53.47	54.00
P.ct. of CO_2 absorbed by KOH	6.40	7.49	8.00
Difference to make up 100 p.ct.	3.49	4.82	3.34
Volume of gas after absorbing CO_2	4,000 c.c.	5,000 c.c.	4,000 c.c.
P.ct. composition of this gas — CO	92.25	87.36	84.61
O	—	1.11	1.65
CH_4	2.96	4.15	4.32
N by difference	4.89	7.38	9.42
P.ct. of pitch from distillate	7.69	7.49	11.33
P.ct. of acetic acid	5.58	6.02	5.76
P.ct. of methyl alcohol	1.32	5.31	10.75
Maximum temperature	344°	380°	343°

case of jute, by a strongly exothermic reaction occurring in all cases at about 320°.

From the general literature of the subject, which is somewhat scattered, we find that the following compounds have been identified amongst the liquid products:

Aqueous distillate		Tar	
Acids	*Alcohol*	*Aldehydes,*	*Hydrocarbons,*
Chiefly Acetic.	Chiefly Methyl	*Ketones, &c.*	*Phenols, &c.*
Also Formic		Acetaldehyde	Paraffins
Propionic		Furfuraldehyde	Toluene
Butyric		Acetone	Xylene
Valerianic, &c.		Methyl-propyl ketone	Creosol
Crotonic acid		Methyl ethyl ,,	Guaiacol
		Methyl formate	Methoxy-derivatives of pyrogallol
		Methyl acetate	Methyl-pyrogallol
			Propyl-pyrogallol

A large number of these derivatives are obviously formed by secondary reactions. What chiefly concerns our subject is to distinguish, if possible, the primary products of the decomposition.

A careful survey of the evidence leaves no doubt that the main products, as to relative quantity, are the primary products, viz.:

Methyl alcohol   Acetic acid   Furfural and Pyrogallol derivatives

In regard to the two former, interesting conclusions are drawn by Ramsay and Chorley from their results (*loc. cit.*). It will be noted that these products are constant for the individual woods, over a wide range of temperatures of distillation, and it is probable that they are formed as it were explosively, i.e. in the exothermic reaction above described.

In regard to the relationship of furfural and acetic acid to each other, the following results of observations upon a particular distillation of alder wood may be cited.

No.	Temperature	Time	Quantity of distillate	Furfural	Acetic acid
		Hours	Less than	Grm.	Grm.
1	200°	2	1 c.c.	0.05	0.0
			c.c.		
2	200-230°	2	4.5	0.09	1.28
3	230-250°	1	6.0	0.096	1.95
		Min.			
4	250-270°	25	4.5	0.05	0.88
5	270-290°	25	4.5	0.064	0.83
6	290-310°	15	3.5	0.086	0.80
7	310-320°	10	5.0	0.142	0.79
8	320-330°	10	5.0	0.170	0.78
9	330-340°	15	4.5	0.237	0.87
		Totals . . .		0.985	8.18

Both are formed continuously, and increase towards the end of the distillation. Special observations were also made on the yield of acetic acid from oak (183 grms. of wood), with the following results:

No.	—	Yield	Containing acetic acid
		C c.	Grms.
1	Up to 120°	20	—
2	120-180°	10	0.1756
3	180-240°	10	0.7320
4	240-260°	10	1.0248
5	260-300°	10	1.4640
6	300-310°	10	1.5372
7	310-322°	10	1.0980
8	322-350°	15	2.3424
9	350-450°	About 10	1.6104

On comparing the woods with one another the only very noteworthy feature of difference is the very large yield of methyl spirit obtained from alder wood. The acetic acid results are approximately constant for this series of woods.

In regard to the aromatic products it is important to note the predominance of pyrogallol derivatives, and that many of these are characterised by the OCH_3 group.

This strongly confirms the view we have taken of the constitution of the hexene constituents of the lignocelluloses. The completion of the benzene ring under the conditions of the distillation indicates its occurrence in the ordinary life-history of the wood. Unfortunately the mechanism of the condensation is too complex to follow, and so, in fact, of the entire process. What is required is an extended investigation of this destructive decomposition under the conditions of variations determined by the addition of reagents, added to promote reaction in one or other direction. All that we can deduce from the results of investigations as they stand is a general confirmation of previous discussions of the constitutional relationships of the constituent groups of the lignocelluloses.

(2) PROCESSES OF DISINTEGRATION BY REAGENTS.—A. *Proximate resolutions.*—The various processes of preparing a papermaker's pulp from the woods admit of a simple theoretical classification on the basis of the foregoing treatment of the subject. The table on p. 209, from a paper of the authors (Forestry Exhibition Reports, Edinburgh, 1886, No. 20), gives such a comparative survey, together with the names of the inventors more prominently associated with the origination of the several methods.

The general principles of the classification are briefly these : The lignocelluloses are readily attacked by hydrolysing agents, even water. The attack of these agents is accompanied by the inverse processes of condensation, which may be and are manysided, owing to the presence of OH groups of the most varied function. A limit is therefore reached when the product is sufficiently condensed to resist further attack. There are two general ways of extending the limit : (1) strengthening the hydrolysing action, either by concentration of the reagent, or by increase of temperature ; (2) preventing the reverse action by fixing reactive groups in combination. These more active

CLASSIFICATION OF THE CHEMICAL PROCESSES OF DISINTEGRATING WOOD.

AQUEOUS ALKALIS.	WATER.	AQUEOUS ACIDS.	
Hydrolysis aided by alkali directly, also indirectly by combination with products. Reversal of hydrolysis aided by temperature, oxidation, and gradual neutralisation of alkali.	Hydrolysis aided by acids formed at the expense of the wood. Limit determined by reversal of hydrolysis, i.e. dehydration aided by oxidation.	(a) *Oxidising and hydrolytic.* Nitric acid. Nitro-hydrochloric acid.	{ COURTER & MELLIER, 1852. BARRE & BLONDEL, 1861. OKIOLI, 1865.
WATT & BURGESS, 1853. HOUGHTON, 1857.	FRY, 1867.	(b) *Simply hydrolytic.* Hydrochloric acid.	BACHET & MACHARD, 1864.
		(c) *Reducing and hydrolytic.* Sulphurous acid.	{ TILGHMANN, 1866. PICTET, 1852.

SOLUTION OF ALKALINE SULPHIDES.	WATER TOGETHER WITH NEUTRAL SULPHITES.	BISULPHITES.	
Hydrolysis aided by alkaline bases directly and indirectly by combination with products. Reversal of hydrolysis lessened by presence of reducing agent.	Simple hydr. lysis, requiring therefore higher temperature. Products removed from sphere of action by combination with base and with sulphite residue. Oxidation prevented by presence of sulphurous acid.	Hydrolysis aided primarily by sulphurous acid, and secondarily by combination of products with bisulphites, also by prevention of oxidation.	TILGHMANN, 1866 MITSCHERLICH, 1874. EKMAN, 1881. FRANCKE, 1882. GRAHAM, 1882.
JULLION, 1855. BLITZ, 1883. DAHL, 1884.	CROSS, 1880.		

groups are chiefly two, viz. aldehydic and acid; and hence the employment of sulphurous acid and bisulphites on the one hand, and alkalis on the other, for the purposes as indicated in the table.

The application of these principles is well illustrated by taking any of the processes in series of variations. Thus the sulphite processes:

(a) *Sulphurous acid* alone is capable of resolving wood into cellulose (insoluble) and non-cellulose (hydrolysed), and soluble derivatives. The acid, however, owing to its feeble hydrolysing power, requires to be used in 7 p.ct. solution (SO_2) prepared under pressure; and, again, to prevent reverse action, the limit of temperature employed is 105°. (See 'The Pictet-Brelaz Process of Preparing Wood Cellulose,' Cross and Bevan. Spon: London, 1889.)

The hydrolysing action of the $SO_2.Aq$, ordinarily very feeble, is perhaps also more powerful in relation to aldehydic condensations.

(b) *The bisulphites.*—The addition of the base lowers the hydrolysing action of the acid; a higher temperature (150-160°) is therefore required. The base, however, serves to saturate acid groups, and the process is further aided by sulphonation in the $CH=CH$ groups; the presence of the excess of bisulphite prevents reversal by condensation of aldehydic groups.

(c) *Neutral sodium sulphite.*—In this case the hydrolysing action is still feebler, and a higher temperature is required (160–180°).

In presence of the lignocellulose complex, undergoing decomposition, the sulphite is dissociated, the base going to acid groups, and the acid sulphite residue to aldehydic groups.

In all the above processes, moreover, the resolution is aided by deoxidation of the lignocellulose constituents, a certain proportion of sulphate being formed.

The alkaline processes involve reactions of a totally different character. In the sulphite processes the resolution of cellulose from lignone is a comparatively simple process; the latter is obtained as a soluble derivative but little changed in essential chemical features from the condition in which it existed in the wood (see p. 178). The alkalis determine, on the other hand, a highly complex decomposition; the products are extremely numerous, and for the most part ill-defined. An analytical study of these products will be found in the Papier Zeitung, 1878, 226, 242. A prominent feature of the decomposition is the liberation or formation of acid groups, and the consequent 'saturation' of the alkali. The hydrolysing power of the alkaline solution is continually diminished, and the alkali has therefore to be used in excess; and according to the amount taken in excess, so, inversely, is the temperature necessary for completing the decomposition. The additional presence of reducing agents, such as sulphides, appears to have a certain influence upon the result. But since the organic products themselves, in presence of the alkali, are of a powerfully deoxidising character, any influence would probably be traceable to specific reactions between the sulphur and the constituents of the wood.

The acid processes we consider with exclusion of the sulphite processes. They divide themselves into the two groups: (*a*) resolution by non-oxidising acids (dilute H_2SO_4 and HCl); (*b*) by oxidising acids (dilute HNO_3).

In the former we have a process of some theoretical interest, consisting of boiling with HClAq, neutralising, and bringing the solution into alcoholic fermentation. It is stated in the historical notices of the industrial development of the process that it was worked for some time on a commercial scale (Payen, Wagn. Jahresber. 1867). In such treatments the limit of resolution is rapidly attained, and the residue is in the brittle and friable condition of the more 'condensed' products. The limit is obviously determined by the inability of such acids to enter into any permanent synthetical combination with the constituent groups of the wood-substance, which remain therefore open to mutual interaction; hence their further condensation, and the building up of more resistant forms of lignocelluloses. The reaction with nitric acid, on the other hand, is of a totally different order: the acid hydrolyses and oxidises in the first instance, but the specific and characteristic decomposition which ensues (described in detail, p. 146) is the result of direct synthesis of the lower oxides of nitrogen with the lignone groups. The final products of resolution of the non-cellulose, or lignone, are the simplest acids—carbonic, acetic, and oxalic.

The extreme oxidising action of nitric acid has been investigated by Wheeler and Tollens (Annalen, 267, 367) in the case of pine-wood, the wood being heated for 6 hours at 90-100°, with 10 times its weight of nitric acid (1·4 sp.gr.) diluted with $\frac{1}{2}$ water. The lignone constituents are, of course, entirely broken down in the reaction, the greater proportion of the cellulose also. The residue was an oxycellulose, amounting to 17 p.ct. of the weight of the wood, having the composition:

		Calc. $C_{12}H_{20}O_{11}$ $=6C_xH_{10}O_5+O$
C	43·41	43·72
H	6·19	6·07
O	50·40	50·20

Its properties were those of the oxycelluloses obtained by the action of nitric acid upon cotton.

The peculiar feature of this oxidation is the survival of a residue so little different in empirical composition from the wood cellulose itself.

The results of these various treatments confirm entirely the views advanced as to the constitution of the lignocelluloses, and, conversely, a grasp of these views enables us to predict, with a satisfactory approximation, the results of treatments not specifically investigated.

B. *Ultimate resolutions.*—(1) The extreme action of the alkaline hydrates at 200-300° is an industrial process of some importance for the preparation of oxalic acid from waste wood. An exhaustive investigation of the process, more especially of the conditions determining maximum yields of the acid, has been made by W. Thorn (Dingl. J. 210, 24). The optimum temperature for the decomposition is 240°, but at this point a relatively large proportion of the alkali is required for complete decomposition, viz. 4 parts hydroxide to 1 part wood. Potassium gives higher yields than sodium hydroxide : the maxima obtained under the condition of heating in closed vessels were, 52 p.ct. (NaOH) + 66 p.ct. (KOH); heated in thin layers, the yield in the latter case was increased to 80 p.ct.

The maximum yields from various woods observed by the author, and calculated on the *dry* woods, were :

Pine, 94·7 ; Poplar, 93·14 ; Oak, 83·4 ; Box, 86·4.

By a more graduated but still severe treatment with the alkaline hydrates, G. Lange obtains the following characteristic products of resolution :

(1) Cellulose (insoluble), and in the alkaline liquid.

(2) Two complex acids, or groups of acids (lignic acids), described below.

(3) Formic and acetic acids, and traces of the higher fatty acids.

(4) Protocatechuic acid and catechol ; (5) ammonia and nitrogenous bases in small quantity.

The lignic acids are of definite composition so far as is established by uniform results of elementary analysis, which are as follows :

Wood from which obtained	Lignic acid soluble in alcohol		Lignic acid insoluble in alcohol	
	C	H		
Beech	61·5	5·5	59·0	5·4
Ash	61·6	5·5	58·8	5·2
Pine	61·3	5·0	60·5	5·2

Nothing, however, has as yet been established in regard to the constitution of these products.

They are no doubt in the main derived from the non-cellulose or lignone complex, but the author's numbers do not warrant his view that the attack of the alkali is confined to this complex; it is evident that the cellulose is considerably attacked also, and the method cannot be recommended as a process for cellulose estimation. For the original papers see Zeitschr. Physiol. Chem. 14, pp. 15, 217.

(2) *Chromic acid in presence of sulphuric acid* (conc.) determines complete conversion of the carbon of the woods into the gaseous oxides CO_2 and CO; the proportion of the latter is small. The reaction is available, therefore, as a combustion method, under the conditions previously described.

Pectocelluloses and Mucocelluloses.—The second great division of the compound celluloses are those of which the non-cellulose constituents are related to the 'pectic' group of compounds. Hugo Müller (Pflanzenfaser), in stating the results of proximate analyses of raw fibrous materials, completes the list of constituents with an undetermined aggregate described as 'Incrusting and intercellular substance and pectic constituents, calculated from loss'—i.e. having determined ash, water, water extract, fat and wax, and cellulose, the residue is estimated by calculation, and stated under the above aggregate description. The question of 'incrustation' is rather morphological than chemical. In the lignocelluloses we have ample evidence that the process of lignification is an intrinsic transformation of tissue-substance. It is, of course, consistent with this view that there should be a concurrent process of deposition of substance external to the cells themselves, destined for the binding or cementing of the cells together into a compact tissue. But it is important not to confuse effects, even though they may proceed from the same cause. In the lignocelluloses, the view of incrustation of the cells is too often transferred to the chemistry of the cell-substance, which comes to be regarded as composed of cellulose merely overlaid with non-cellulose constituents which mask its reactions. These views, we maintain, should be kept distinct.

The morphology of cell formation teaches that with growth a differentiation of a portion of the cell wall takes place, which in the fully developed condition of the tissue constitutes the division of cell from cell and completes their individualisation. A true 'intercellular' region is then formed, and chemical differentiation no doubt often takes a different course in this region; but it has not yet been established that differentiation of the cell-substance does not take place simultaneously in the same direction though in lesser degree. In the lignocellulose this certainly appears to be the case; in the pectocelluloses, which we are about to consider, the problem is more difficult to investigate by chemical and microchemical observation, owing to the absence of any well-marked reactions of the pectic compounds.

We must now give a brief outline of the chemistry of this group. Their characteristic property is that of yielding gelatinous hydrates, in which they closely resemble the mucilage-yielding constituents of many seeds, fruits, and rhizomes—e.g. linseed, the seed of *Plantago Psyllium*, the roots of *Orchis Morio*, &c., many of the Salvia species, the fruit of the quince (*Cydonia vulgaris*), &c. While, however, the empirical composition of the latter is that of the carbohydrates, viz. $C_nH_{2m}O_m$, the pectic group are distinguished by empirical formulæ with considerably less hydrogen in proportion to the carbon and oxygen, the general approximate formula being

$C_mH_{3m}O_m$. A second feature of distinction is that, while the former yield on hydrolysis hexoses and pentoses, the latter give the series of pectic acids, this distinction corresponding with the higher proportion of oxygen which characterises the group. It is necessary to qualify these conclusions somewhat by pointing out that the empirical formulæ determined by the original investigators of these compounds—viz. Frémy (Ann. Chim. Phys. [3] 24, 5), Chodnew (Ann. Chem. Pharm. 51, 355)—have been called in question by later observers. Thus Reichardt (Archiv d. Pharm. [3] 10, 116) concludes that they are to be regarded as gelatinisable carbohydrates (see Tollens, Kohlenhydrate, p. 243).

On the other hand it will be found that the pectocelluloses differ from the celluloses by increased proportion of oxygen; and their acid character is further shown by their retaining a relatively large proportion of basic mineral constituents (ash).

The general relationships of the group as determined by the earlier observers are these: *Pectose*, the insoluble mother substance of the group, occurs in mixture or union with the cellulose of the parenchyma of fleshy fruits and roots, e.g. apples, pears, turnips, &c. This is hydrolysed by boiling dilute acids or alkalis, or by a ferment enzyme (pectase) secreted in the tissue, to *pectin* ($C_{32}H_{48}O_{32}$, Frémy), the solutions of which readily gelatinise. By continued hydrolysis (boiling water) this is further modified to parapectin, and by alkalis to metapectin and parapectic acid and pectic acid ($C_{32}H_{44}O_{30}$, Frémy; $C_{12}H_{16}O_{11}$, Regnault; $C_{12}H_{16}O_{10}$, Mulder; $C_{14}H_{22}O_{14}$, Chodnew).

The final product of hydrolysis is metapectic acid. To this acid Frémy assigned the formula $C_8H_{14}O_9$. Later investigations have established its general identity with arabic acid—

a complex acid which is the main constituent of gum-arabic. Gum-arabic yields, on graduated hydrolysis, a complex of glucoses (galactose, arabinose) and a series of arabinosic acids, e.g. $C_{23}H_{38}O_{22}$, and compounds differing from this by $+ C_6H_{10}O_5$. It appears, therefore, generally, that the pectic group are compounds of carbohydrates of varied constitution with acid groups of undetermined constitution, associated together to form molecular complexes, more or less homogeneous, but entirely resolved by the continued action of simple hydrolytic agencies; and the *pectocelluloses* are substances of similar character in which the carbohydrates are in part replaced by *non-hydrolysable celluloses*. The general characteristics of the *pectocelluloses* are therefore these: they are resolved by boiling with dilute alkaline solutions into *cellulose* (insoluble) and soluble derivatives of the *non-cellulose* (pectin, pectic acid, metapectic acid); they are gelatinised under the alkaline treatment; they are 'saturated compounds,' not reacting with the halogens, nor containing any groups immediately allied to the aromatic series.

Compound celluloses of this kind are enormously diversified in composition, structural character, and distribution, and the group, having none of the sharp lines of differentiation and demarcation presented by the lignocelluloses, cannot be handled at all in the same way.

We must confine ourselves, therefore, to the one or two more definite types which have been investigated.

Flax.—Commercial flax is a mixed product. The bast fibre proper constitutes from 20–25 p.ct. of the entire stem, and is more or less imperfectly separated from the wood on the one side, and the cortical tissue-elements on the other, by the ordinary processes of retting and scutching. These residues are visible with the naked eye, but are brought into clearer evidence by means of reagents, followed by microscopic examination.

Thus the *wood* is an ordinary lignocellulose, and gives the characteristic reactions; the *cortical tissue* is again distinguished from the fibre proper by reacting strongly with magenta-sulphurous acid. The presence of the cortical tissue is also marked by the large proportion of 'oil and wax' constituents present in the fibre (3-4 p.ct.). Excluding these adventitious constituents the fibre proper is a pectocellulose. That the non-cellulose constituents of flax are pectic compounds was first established by Kolb (Bull. Soc. Ind. Mulhouse, June 1868). According to his observations, the precipitate obtained on acidifying the alkaline solutions from the 'boiling' of flax goods consists of pectic acid.

The proportion of these constituents varies from 14-33 p.ct. in the different kinds of flax, the variations being in part due to the plant, i.e. to physiological habit and conditions of growth; in part to the different methods of retting the plant— and extracting the fibre. After well boiling with the dilute alkali (1-2 p.ct. NaOH) the fibre-substance consists of flax cellulose, with residues of the wood (sprit), cuticular tissues, and oils and waxes associated with the latter. By exposure to chlorine (after well washing and squeezing) the wood is attacked in the usual way, and is then easily resolved by alkaline treatment. To purify the cellulose it requires to be boiled out with alcohol, and finally treated with ether-alcohol to remove the oil-wax residues. In this way flax cellulose is isolated in the laboratory in an approximately pure condition. It might appear from the outlines of this laboratory method that the bleaching of flax goods, which consists substantially in the isolation of the pure flax cellulose, is a comparatively simple process. This is not so, however. The exigencies of economical and safe treatment of textile fabrics prescribe certain narrow limits of chemical treatment; and the removal of the more resistant wood (lignocellulose) and cuticle (adipocellulose) under these conditions involves a reiterated round of treatments consisting of—

Alkaline hydrolysis	Boiling in solutions of NaOH, Na_2CO_3, &c.
Oxidation	Hypochlorite solutions and atmospheric oxidation (grassing).
Souring	Treatment with dilute acids in the cold.

It must be remembered, however, that the problem is not the removal of the non-cellulose constituents of the fibre itself—these disappear almost entirely in the earliest alkaline treatments—but of compound celluloses of the other two main groups.

The further investigation of the pectose of flax fibre has not been prosecuted according to the methods of later years. Such investigations will, no doubt, be undertaken in due course.

FLAX CELLULOSE has been mentioned incidentally to the general treatment of the celluloses. So far no reactions have been brought to light in which it is differentiated from cotton cellulose, with perhaps one exception, viz. its lesser resistance to hydrolysis. Thus H. Müller mentions (Pflanzenfaser, p. 38) that flax cellulose isolated by the bromine method lost, on boiling five times with a dilute solution of sodium carbonate (1 p.ct. Na_2CO_3), 10 p.ct. of its weight. The statements of R. Godeffroy (abstracted in J. Soc. Chem. Ind. 1889, 575), that flax cellulose is distinguished from cotton cellulose by its reducing action upon silver nitrate in boiling neutral solution, are erroneous, the reaction resulting from residual impurities, which, for the reasons given, are extremely difficult to isolate. Flax cellulose may therefore, for the present, be regarded as chemically indistinguishable from cotton cellulose.

The oil and wax constituents of the raw fibre will be described under the group of *adipocelluloses*.

OTHER PECTOCELLULOSES.—As far as investigation has proceeded, it appears that pectose, or pectose-like substances, are associated with all fibrous tissues of the unlignified order.

And indeed in the lignocelluloses themselves pectous substances make their appearance with increasing age. Thus the lower portions of the isolated jute bast—jute cuttings or butts —when boiled in alkaline solution yield products which cause the solution to gelatinise on cooling ; and the gelatinous product is insoluble in alcohol, distinguishing it, as pectic acid, from the products of hydrolysis of the lignocellulose itself, which are dissolved, after precipitation, by alcohol. It must be remembered, however, that in the 'jute cuttings' the adhesion of the bark and cortical parenchyma to the true bast fibre is such that we are dealing with a complex tissue, and the source of the pectic acid may be in the parenchyma of the tissue and not in the bast fibre. On the other hand, we have shown (p. 152) that in the spontaneous decomposition of jute, lying in the damp state, gelatinous acid bodies are formed, indistinguishable from pectic acid. It would not be difficult, therefore, to account for the pectic constituents of the bast tissue towards the root end, as products of degradation of the lignocellulose itself.

Reverting, however, to the non-lignified fibres such as China grass, or Ramie (Böhmeria species), and the 'nettle fibres' generally, hemp, and even raw cotton—these all contain pectic bodies associated with the cellulose, which are hydrolysed and dissolved by treatment with boiling alkalis. But these pectocelluloses have not been sufficiently investigated *as* compound celluloses to admit of any useful classification on the basis of particular constitutional variations of their non-cellulose constituents.

The monocotyledonous fibre-aggregates, whether fibrovascular bundles (Phormium, Aloe fibres, Musa, &c.) or entire plants (Esparto, Bamboo stems, Sugar Cane), are largely made up of pectocelluloses, with a greater or less proportion of lignocelluloses. But the constitution of these non-cellulose constituents is as yet quite unknown, and we have therefore none but the general basis of classification.

In the same way also the parenchymatous tissue of fruits, fleshy roots, &c.—the typical pectocelluloses—must be, for the present, dismissed with the bare mention.

The investigation of these substances belongs rather to the province of general carbohydrate chemistry than to the narrower cellulose group; and the problems involved are in many respects rather morphological and physiological than purely chemical.

These same considerations apply also in great measure to the mucilaginous constituents of plant tissues, though certain of these have been investigated by modern chemical methods. The relationship of these substances to cellulose is indicated (*a*) by the histology of the tissues, which shows them to be associated with the *cell wall*, rather than with the cell contents; (*b*) by their empirical composition, which is approximately that of cellulose; (*c*) by their reactions with iodine, by which they are coloured variously from blue to violet, as are the hydrated modifications of cellulose (Sachsse, Farbstoffe, &c. p. 161). Beyond superficial observations of reactions (iodine) and gelatinisation with water, these compound celluloses—which may conveniently be termed mucocelluloses—had been but little investigated (Sachsse, *loc. cit.*) until the systematic work of Kirchner and Tollens, and Gaus and Tollens (Annalen, 175, 205; 249, 245), upon the mucilages and gums. Of these typical researches we give a brief account.

(1) QUINCE MUCILAGE was prepared by digesting 50 grms. with 1 litre of warm water, pouring off, and repeating the digestion; filtering the mucilage by squeezing through cotton cloth, and precipitating the dissolved product by the addition of hydrochloric acid and alcohol. After washing with alcohol and ether the product was dried, forming brittle fibrous masses;

the yield amounted to 8-10 p.ct. Ultimate analysis of the products, retaining 5-6 p.ct. inorganic constituents, gave the following numbers, varying between the extremes

$$
\begin{array}{lcc}
\text{C} & 46\cdot52 & 44\cdot17 \\
\text{H} & 5\cdot88 & 6\cdot15 \\
\end{array}
$$
Corresponding to the formula $C_{18}H_{28}O_{11}$ $C_{18}H_{30}O_{15}$

The product was then investigated for the presence of typical carbohydrate groups.

Oxidation with nitric acid gave no mucic acid and no saccharic acid. *Galactose* and *dextrose* groups are therefore absent. On the other hand, *furfural* was obtained in some quantity (6·45 p.ct. furfuramide) on distillation with acids. The substance therefore contains pentose groups.

Hydrolysis with dilute sulphuric acid.—On boiling with the acid, the product is resolved into—

Cellulose	*Gummy bodies*	and	*Glucoses*
Insoluble and amounting to 23 p.ct.	Precipitated by alcohol from neutralised solution		Soluble in alcohol

And a mixture of

From the soluble products it was found impossible to isolate any glucose in the crystalline form. The solution, on the other hand, certainly contained compounds of this group, since it was strongly dextro-rotary, reduced Fehling's solution to an amount equal to 62 p.ct. that of dextrose, and gave, with phenylhydrazine, an osazone melting at 162°, and giving results on analysis corresponding with a mixture of osazones of a pentose and hexose.

It is evident from these results that the mucilage is comparatively resistant to hydrolysis; by its behaviour, in fact, it is shown to be much more nearly related to the cellulosic than the starch type of 'saccharo-colloids.' It is for this reason that we direct special attention to this remarkable group of compounds, since their further investigation cannot fail to throw light upon the problems discussed in the earlier sections of this treatise.

(2) SALEP MUCILAGE was prepared from the tubers previously pulped by grinding in a mortar, the details of preparation being exactly as for the above. The mucilage was precipitated by alcohol in white threads which hardened under further treatment with alcohol (dehydration).

The dry substance (retaining 1·5 p.ct. mineral constituents) gave on analysis numbers approximately those of cellulose ($C_6H_{10}O_5$), viz. :

$$C \quad . \quad . \quad . \quad . \quad 44\cdot58$$
$$H \quad . \quad . \quad . \quad . \quad 6\cdot63$$

The hydrolysis of the product, by boiling with dilute acid (1·25 p.ct. H_2SO_4), was investigated in relation to the influence of the time factor upon the three products, cellulose, gum, and glucoses: the former being estimated by direct weighing; the latter, in terms of dextrose, by titration with Fehling's solution; the result, subtracted from the total dissolved products, giving the yield of gum.

Duration of hydrolysis	Cellulose	Gum	Glucose
¼ hour	16·84	—	11·46
1 ,,	14·93	49·33	41·93
2 hours	—	44·92	53·29
3 ,,	11·58	29·91	71·27
4 ,,	12·76	18·70	81·37
5 ,,	12·41	16·02	75·97
7 ,,	9·04	6·87	74·76

It will be noted that the sum of the percentages in some cases exceeds 100, and in some is in defect.

These observations are explained by the attendant phenomena of hydration and dehydration; and the disappearance of 'glucose' after the fourth hour, when it reaches a maximum, is evidently due to condensation of aldehydic groups.

The further investigation of the product established in this case the *absence* of *galactose* and *pentose* groups, but the presence of *dextrose* groups in small proportion. The product of hydrolysis yielded a mixture of osazones in which the derivatives of dextrose and mannose appear to be represented.

But again the constitution of the products of hydrolysis is left in a state of incomplete elucidation. That the authors' methods failed to solve these problems further than has been shown is a further illustration of the complexity of the subject. It is, in fact, the expression of the difficulty invariably experienced with products of the cellulose class, viz. resistance to hydrolysis; and it is from the internal evidence of the difficulties experienced by such practised investigators that we are the more inclined to regard these products, although soluble in water, as cellulose derivatives.

In regard to other mucilages, we may briefly mention the more important, in order to give some idea of the distribution of these compounds in the plant world.

AMYLOID is the name applied to a mucilaginous product obtained from the seeds of a number of the Leguminosæ, e.g. *Tamarindus indica*, *Hymenæa Courbaril*, and *Schotia latifolia* (Schleiden, Beiträge z. Botanik, i. 168). It is soluble in boiling water, partly also in cold. It is precipitated by an alcoholic solution of iodine as a blue flocculent mass.

A similar substance was obtained by Frank (Pringsheim, Jahrb. f. Wiss. Bot. 5, 15) from the membranes of the cotyledon cells of *Tropæolum majus*. This product was also definitely proved to be formed at the expense of starch.

LICHENIN is the soluble constituent of the membranes of *Cetraria islandica* ('Iceland moss') and other similar lichens (Knop and Schnedermann, Annalen, 55, 164). It is extracted by treating the plant product with cold dilute hydrochloric acid and adding alcohol to the solution; or by boiling with water, after previously purifying the raw material by digesting with dilute alkaline solutions in the cold. According to Hönig and Schubert (Wien. Akad. Ber. 96, [2] 685), lichenin is accompanied in the plant by an amorphous form of starch.

On hydrolysis lichenin yields crystallisable dextrose, and on oxidation with nitric acid, saccharic acid. With glacial acetic acid it yields an amorphous triacetate, $C_6H_7O_2(C_2H_3O_2)_3$.

CARRAGHEEN MUCILAGE is obtained from the seaweed *Fucus crispus* (C. Schmidt, Annalen, 51, 56) on boiling with water. This raw material is characterised by the presence of galactose groups, yielding 22 p.ct. mucic acid on oxidation with nitric acid, and also crystallisable galactose on hydrolysis with boiling dilute acids.

This concludes our brief notice of the pectocellulose group. It appears that there are two well-marked sub-groups of these products: (1) the pectocelluloses proper, occurring in structures of a more permanent character—fibrous and parenchymatous tissues of the stems and roots of Phanerogams; (2) pectocelluloses occurring chiefly in seeds and fruits of Phanerogams and the tissues of Algæ; distinguished from the former by yielding to the action of water, giving the peculiar solutions known as mucilages. Hence the proposed name mucocelluloses for the parent tissue-substance having these properties.

These groups are further distinguished by the characteristics of their hydrolysable constituents, the former yielding complexes in which acid features predominate; the latter yield neutral solutions, and in fact, on ultimate hydrolysis, various hexoses and pentoses.

Adipocelluloses and Cutocelluloses.—Cork and Cuticularised Tissues.—The plant represents, in the one view, an assemblage of synthetical operations carried on within a space enclosed and protected from the destructive influences of water and unlimited atmospheric oxygen. The protecting external tissues are those which we are about to describe as constituting the third important group of compound celluloses. These tissues contain, in admixture with the tissue-substance, a variety of oily and waxy products (easily removed by mechanical solvents), the presence of which adds very considerably to the water-resisting property of the tissue. It will be seen as we proceed, however, that the tissue-substance, after being entirely freed from these adventitious constituents or oily excreta, yields a large additional quantity of such products when decomposed by 'artificial' processes of oxidation and saponification. By this and by its empirical composition (*infra*) the tissue-substance will be seen to contain 'residues' of high carbon percentage and molecular weight, and closely allied in chemical structure to the oil and wax compounds found in the 'free' state in the tissue as it occurs in the plant. These groups are associated in combination in the tissue with cellulose residues, and hence the description of such complexes as adipocelluloses.

CORK in its ordinary form is a complex mixture containing not only oils and waxes, but tannins, lignocelluloses, and nitrogenous residues. The following are the results of elementary analysis : (*a*) of cork purified by exhaustive treatment with ether, alcohol, and water ; (*b*) of cork (*Quercus suber*) without purification ; (*c*) of the cork tissue of the cuticle of the potato (tuber) purified by exhaustion with alcohol.

	(a)	(b)	(c)
C	67·8	65·7	62·3
H	8·7	8·3	7·1
O	21·2	24·5	27·6
N	2·3	1·5	3·0

The analyses are calculated on the ash-free substance (Döpping, Annalen, 45, 286 ; Mitscherlich, Annalen, 75, 305).

These investigators succeeded in isolating cellulose from cork, but by complicated and drastic methods of treatment, such as would break down the greater proportion of the cellulose into soluble derivatives. These treatments were : (1) drastic oxidation with nitric acid ; (2) alternate treatments with boiling dilute hydrochloric acid and 10 p.ct. solution of potassium hydrate. The proportion thus isolated amounted to 2-3 p.ct. only.

The authors, on the other hand, have observed that the non-cellulose of cork is entirely converted into soluble derivatives by the process of digestion at high temperatures with solutions of the alkaline sulphites as described, p. 150. In this way a residue is obtained preserving the form, i.e. cellular structure, of the original cork, and amounting to 9-12 p.ct.

The details of a particular experiment were as follows : 10·995 grms. cork, 20 grms. $Na_2SO_3.7H_2O$, 2 grms. Na_2CO_3, dissolved in 500 c.c. water. Digested 3 hours at 75 lbs., and 4 hours at 125 lbs. pressure. Residue bleached with sodium hypochlorite solution. Yield of cellulose, 1·34 grms. ; 12·1 p.ct.

In the proximate analysis of cork M. Siewert found 10 p.ct. of constituents soluble in alcohol, which were further resolved into—

Wax, in crystalline form	1·75
Fat acid, non-crystallisable	2·50
Acid (2), non-crystallisable and of fatty character	2·25
Tannic acid, soluble in water	2·50
,, ,, soluble with difficulty	1·00

The crystallised wax is termed by Siewert phellyl alcohol, $C_{17}H_{28}O$. It melts at 100°, and dissolves in 500 parts boiling alcohol. The acid bodies are described as (1) decacrylic acid, $C_{10}H_{18}O_2$ (m.p. 86°), soluble in 52 parts boiling alcohol ; (2) eulysin, $C_{24}H_{36}O_3$ (m.p. 150°), soluble in cold alcohol.

According to Höhnel (Wien. Akad. Ber. 76) and Kugler (Dissert. on Suberin, Halle, 1884), the cork-substance proper is a mixture of cellulose with lignocellulose and two characteristic compounds, cerin and suberin. Cerin has the empirical formula $C_{20}H_{32}O$. Suberin is of a fatty nature, and yields stearic acid and phellonic acid ($C_{22}H_{42}O_4$) on saponification.

In regard to the nomenclature of these compounds, we may, consistently with the plan of this treatise, adopt the terms Suberose and Cutose for the compound adipocelluloses as they occur in the plant, and Suberin and Cutin for the non-cellulose groups united to the cellulose to form the entire complex.

This complex of substances has been further investigated by Flückiger (Arch. Pharm. 228, 690), who obtains glycerol as a product of saponification, showing the presence of the more ordinary glycerides of the fat acids; he otherwise confirms, in the main, the results of Kugler (*supra*). Still more recently by C. v. Wissenburgh (Chem. Centr. 1892, ii. 516); but the later results throw no further light on the more important aspects of the problem. These are obviously the questions of the constitution of the main tissue-substance and the physiology of its formation, as well as that of the waxy excreted products which accompany it.

In suberose, as in cutose, it is to be noted that Frémy overlooks the cellulose residue. V. Wissenburgh (*loc. cit.*) makes the more positive statement that cork contains no cellulose. Both observers appear to be in error on this point.

There is some more conclusive evidence on these points in the earlier work of Döpping and Mitscherlich. It is found in effect that when rasped cork—yielding to solvents 7-10 p.ct. of fatty constituents (*supra*)—is oxidised with nitric acid, it yields 40 p.ct. of fatty acids, and acids identical with those obtained from the oxidation of fats and oils under similar conditions, chiefly suberic acid. It is evident, therefore, that the cork tissue is largely made up of constituents standing in very close constitutional relationship to the natural fats and oils, though possessing very different physical properties.

These relationships have been more definitely made out by Frémy, in his investigations of the closely allied compound which constitutes the epidermal or cuticular tissue of the leaves, stems, &c., of Phanerogams. As the characteristic constituent of cork is termed suberin, so Frémy terms this cuticular tissue-substance cutin or cutose. To prepare this substance, the cuticular tissue ('peel') of the apple, e.g., is treated with boiling dilute acids, followed by digestion with the cuprammonium reagent (p. 10); then again with boiling acid, and dilute alkali (KOH); finally the residue is treated with alcohol and ether. In this way a nitrogen-free residue is obtained having the empirical composition—

$$\begin{aligned} C &\quad . \quad . \quad . \quad 73\cdot 66 \text{ p.ct.} \\ H &\quad . \quad . \quad . \quad 11\cdot 37 \text{ ,,} \\ O &\quad . \quad . \quad . \quad 14\cdot 97 \text{ ,,} \end{aligned}$$

Not only by these results, but by the study of the proximate resolutions of this substance it is shown to have the closest relationships to the carbon compounds of the 'wax' class (Compt. Rend. 48, 667).

In a later investigation of the products of saponification of this substance, Frémy worked upon a raw material similarly prepared, but having the composition C 68·3, H 8·9, O 22·8. This compound is termed *cutose*, in substitution for cutin. Cutose is slowly attacked by boiling alkaline solutions; a product is dissolved, *of the same empirical composition* as cutose (comp. Lignocelluloses, p. 157), but of a fatty nature. It is precipitated on acidifying the solution; the precipitate is soluble in ether-alcohol, and when isolated is found to melt below 100°. Under more drastic treatment with alkaline solutions the dissolved products are found to be a mixture. Precipitated by acids and treated with boiling alcohol the mixture dissolves; on cooling, the solution deposits an acid (m.p. 85°) in yellow-coloured flocks, which after fusion form a brownish translucent friable mass. This is a compound of *stearocutic acid* and *oleocutic acid* (*infra*), into which it is resolved by further treatment with alkali. The alcoholic filtrate from the solid acid when evaporated gives a viscous residue, of an acid body, *oleocutic acid*.

By the further action of very concentrated potash solution in the yellow acid, stearocutic acid is formed. The potash salt of this acid is white and translucent, insoluble in water and cold alcohol, soluble in boiling alcohol. The free acid (m.p. 76°) is also insoluble in cold alcohol, slightly only on boiling, but dissolves in benzene and in acetic acid on warming. The acid also dissolves freely in alcohol in presence of oleocutic acid. A similar result is seen with the potassium salt, which, though insoluble in water, dissolves in an aqueous solution of potassium oleocutate.

The composition of these two acids is as under:

Stearocutic Acid.

C .	75·00
H .	10·71
O .	14·28

which is expressed by the formula $C_{28}H_{48}O_4$. This formula is confirmed by the analysis of the salts of the acid.

Oleocu ic Acid.

C .	66·66
H .	7·91
O .	25·42

expressed by the formula $C_{14}H_{20}O_4$.

Cutose is regarded by Frémy as a complex of these two compounds, in the proportion of 1 mol. stearocutic acid : 5 mols. oleocutic acid. These acids undergo alteration on heating at 100° in presence of water, passing into insoluble modifications of higher melting point. The original molecular condition, however, is restored on heating with alkaline solutions.

Frémy also made observations upon suberin (or suberose), which yielded similar products of saponification. He therefore concluded that the two products are substantially identical. The products of oxidation by nitric acid are also indistinguishable, viz. chiefly suberic and succinic acids.

These results suggest, from the purely chemical standpoint, that the cellulose of the tissue and the waxy products of excretion stand to one another in a genetic relationship, and the cutose or suberose occupies an intermediate position. The question of a direct conversion of cellulose into wax taking place in these cuticular tissues was definitely raised and discussed by De Bary in his investigations of this group of plant constituents (Bot. Ztg. 1871). It appears from these researches that wax-alcohols are certainly not contained in the cell-sap or protoplasm, and that their origin must be in the cuticular tissues themselves; but the parent substance may be either cellulose, or some compound built up with it in the ordinary course of elaboration. This question is left for the present undetermined. It should be borne in mind, on the other hand, that we have a great number of direct observations upon the physiological equivalence of the carbohydrates and the fats, both in the animal and vegetable worlds; and although the mechanism of the transformation of the one into the other group of compounds remains unelucidated, it is after all not more difficult to imagine than the condensation to furfural. It is, however, not the purpose of this treatise to carry discussion into purely speculative regions; and it is sufficient to state the conclusion that there is ample ground for adopting as a working hypothesis that carbohydrates, or possibly cellulose, are transformed into cutose or suberose, and these, again, into free waxy bodies of lower molecular weight, the whole process representing the change known as cuticularisation or suberisation.

In all the researches above described there is no attempt to travel outside the empirical region of preparing products of decomposition answering to the more obvious criteria of definite composition. Further investigation will need to attack the more important constitutional problems presented by these products. Some preliminary observations in this direction have indeed been made.

Thus on p. 190 we have cited Benedikt and Bamberger's determinations of the methyl number of cork (*Quercus suber*), the percentage of CH_3 as $O.CH_3$ being 2·45, which is approximately the mean number determined for the woods. The authors have made determinations of furfural, with results as under :

Cork	4·5 p.ct.
Cuticle of apple (purified) . .	3·5 ,,

Also of the yield of acetic and oxalic acids resulting from fusion with alkaline hydrate (3 parts by weight NaOH), the fusion being completed at 280°, with results as under :

Acetic acid	6·0 p.ct.
Oxalic acid	2·1 ,,

These, however, are aggregate results, and cannot be specifically referred to the particular constituents of the cork, viz. cellulose, lignocellulose, the suberin constituents proper, or the 'free' waxy constituents. They indicate, however, the directions of research in which results will be obtained, complementary to the work of Frémy, on the more characteristic products of decomposition of the adipocelluloses.

In another direction also results have been obtained throwing some light on the constitution of these complexes, viz. in investigations by Hodges (R. Irish Acad. Proc. 3, 460) and the authors (J. Chem. Soc. 57, 196) of the cuticular constituents of flax.

The flax was treated, in the form of yarn, with boiling alcohol ; a bright (chlorophyll) green solution was obtained, from which, on cooling, a flocculent magma separated. This was filtered, and washed with alcohol, and on treatment with hot water melted to a greenish resinous mass (product A).

The alcoholic filtrate on distillation gave a brownish-green pasty residue (product B).

These products, examined for nitrogen and mineral matter, gave the following results :

—	Nitrogen	Mineral matter	Phosphoric acid P_2O_5 p.ct. of ash
A	0·09	1·7	7·0
B	0·29	1·1	18·4

These numbers are explained by the presence of residues of chlorophyll.

The complex A yielded, on saponification with alcoholic potash a large proportion of ceryl alcohol ($C_{27}H_{35}OH$), which was proved to exist in the original substance, in great measure in the uncombined state ; and at the same time a mixture of oily ketones, the solutions of which were dichroic (orange green). These ketonic bodies were found to be soluble in sodium acetate solution, reacting in this solution with phenyl-hydrazine (acetate), yielding crystalline precipitates.

The residue from saponification of A, after removing these compounds, was an inert resinous substance.

By regulated fusion with sodium hydrate it was decomposed, with formation of a mixture of acid bodies from which cerotic acid ($C_{27}H_{54}O_2$) was isolated. The complex B contained (*a*) carbohydrate derivatives soluble in water, and giving with boiling hydrochloric acid a copious yield of furfural ; (*b*) ceryl alcohol (about 10 p.ct.) and (*c*) the dichroic ketones (about 18 p.ct.) above described ; (*d*) acid bodies (50-60 p.ct.) of a fatty character, giving soluble alkaline salts and insoluble salts with the alkaline earth metals, insoluble also in alcohol.

In regard to the localisation of these compounds in the flax fibre, it is evident that they are associated with the cuticular tissue, which inference is directly confirmed by analyses of a waste product of the spinning mill ('preparing' process) known as 'hackler's' dust. Hackling is a process of combing the fibre, and the dust which accumulates in the treatment, when examined microscopically, is seen to consist, for the most part, of the cortical parenchyma and cuticular tissue. A specimen of this product was found to contain 7·3 p.ct. mineral constituents (ash) and 14·5 p.ct. moisture. Alcohol and ether extracted 11·7 p.ct. of its weight (15 p.ct. calculated on the dry, ash-free substance), the extract having the same general characters as that obtained from the flax itself. Extracted with petroleum ether, the proportion dissolved amounted to 8·4 p ct. (10 p.ct. on the dry, ash-free substance). Estimations of nitrogen in the original substance gave (1) 1·8 p.ct., (2) 2·1 p.ct., of which one-sixth existed in the form of ammonia or amido-compounds.

Lastly, in reference to this complex of cuticular by-products, it must be remembered that the flax plant is subjected to the retting or rot-steeping process, as a preliminary to the mechanical treatments for separating the fibre (bast) from the stem. The characteristics of the resulting spontaneous fermentation are those of *butyric fermentations*. In such decompositions, of course, the wax-alcohols and ethers can take no part, but the oily products of lower molecular weight (acids and ketones) are no doubt in part formed as products of decomposition of a parent substance susceptible of this species of hydrolysis. The entire subject requires extended investigation. Systematic research would, of course, localise the substance—whether cellular tissue or cell contents—undergoing this particular decomposition, and the result would throw direct light upon the origin of fatty substances in the normal living processes. The importance of these constituents as auxiliaries to the spinning qualities of the fibre lends the additional technical interest to the results of such researches; and more generally the special functions of the adipocelluloses in protecting the tissues beneath them from penetration by water point to the desirability of adapting the artificial processes of waterproofing (cellulose) textiles to the lines laid down by the natural process.

These observations bring us to the close of a brief notice of this third group of compound celluloses. The brevity of the treatment is the expression of the very small amount of research which has been devoted to the subject. There are many scattered references to particular products isolated from vegetable tissues by treatment with 'fat and wax' solvents. But a description of these falls outside the scope of the present treatise, which is confined to the treatment of tissue-constituents, with such incidental references to adventitious or excreted products as appear to be genetically connected with the tissue in which they are found.

From this point of view it appears to be established, by similarity of constitution, that the oils and waxes found in the free state in association with the adipocelluloses are closely related to the non-cellulose constituents of the latter; and that they are either degradation products of the latter, or both have a common origin in some anterior form. It is this important physiological problem which future researches must solve.

General View of the Cellulose Group.—The elaboration of the compounds which constitute the subject-matter of this treatise is, in point of mass-effect, by far the main work of the vegetable kingdom. The functions of these compounds in plant life are obviously in part structural or mechanical, in part chemical. The cellular tissue of the plant being regarded as the seat of the essential vital operations of elaboration and metabolism, the fibrous and vascular systems—in addition to taking their specialised part in the general distribution of nutritive and other matters—are the strengthening elements, whereas the cuticular tissues are mainly concerned in closing off and protecting the tissues beneath from the unregulated action of water and air. In regard to the chemical relationship of the tissue-substances to the vital processes by which they are formed, and to which they in turn contribute, we have little but indirect evidence and conjecture to go upon. So far as this evidence is of a purely chemical nature, it has been dealt with in the preceding sections. But it depends in great measure upon the results of investigation by physiological methods, and for such results the special treatises must be consulted. We have to deal, in conclusion, with those changes of the tissue-substances, celluloses and compound celluloses which accompany or follow the cessation of vital activity.

The term 'death' is perhaps more difficult to apply to the vegetable than to the animal organism. In a perennial plant the active life, in the sense of the elaboration of new material, is bound up with portions only of the structure. In an ordinary forest tree, for instance, the leaves are these active agents; the trunk or stem tissues, on the other hand, are largely depleted of the organic nitrogenous matter (protoplasm) upon which the vitality depends; they have ceased to live in the full sense of the term, but, on the other hand, they are known by casual observation to live, in the sense opposed to decay. There are, therefore, various phases of life in the plant recognised by ordinary observation, and more exactly defined by the physiologist; but these phases graduate by insensible stages into the region where decay and chemical disintegration are predominant. If, therefore, it is difficult for the physiologist to draw the line between the life and death of a cell, it is still more so for the chemist, who deals with the cell-substance. Another phase or aspect of the vital process, of which also the science of to-day gives a very slender account, is that of the organic connection of the vast assemblage of cells which constitute a vegetable organism—more particularly the subordination of each unit to the general life-history of the plant, and the extraordinary complex of adjustments which this involves. These, again, are hardly problems for the chemist, save, perhaps, in regard to the organic relationship of the cell-wall to the protoplasm by which it is formed.[1] The most striking general feature of the cellulosic group is that they are *non-nitrogenous* ;[2] it might be reasoned, therefore, that they are, from the first, excreta, and never live in the full sense of the word. Without pushing these considerations into an argument of doubtful value, we may conclude that from the moment of origin the history of the cellulose group is one of progressive withdrawal from the region of the vital processes proper, and that to again enter that region they must undergo a process of proximate resolution as a result of action from without. This reabsorption

[1] The views obtaining on this point are summed up by Goodale at p. 218 of 'Physiological Botany,' under the heading 'The Relations of the Cell Wall to Protoplasm.' They are, in the main, two: (1) The cell wall is formed, by the solidification upon the exterior of a protoplasmic mass, of matters previously dissolved in it, the pellicle thus formed being regarded as an excretion or secretion. This view is held by Hofmeister (Pflanzenzelle) and Sachs. (2) The cell wall is directly produced by conversion of the outer film of protoplasm into cellulose, with which other groups may be mixed or combined (Schmitz, Sitz. d. Niederrh. Ges. Bonn, 1880). This conclusion is based upon the observation that the volume of protoplasm in a cell decreases *pari passu* with increase of the cell wall, and upon the phenomena which attend thickening of the cell wall.

The distinction between these theories, however, is rather morphological than chemical ; and the view expressed in the text may be taken as consistent with either theory of the actual mode of formation of the cell wall.

[2] We are not aware of any specific proof of the assumption that the celluloses are *ab initio* non-nitrogenous. It is possible that they contain NH_2 residues in the earlier phases of growth.

of cellulosic tissues is a frequent phenomenon; the process by which the tissue is broken down is of the character of an ordinary hydrolysis, i.e. is determined by enzymes, and postulates therefore a cellulose susceptible of molecular disaggregation. The 'hemi-celluloses' (p. 87) are compounds of this order, and occur chiefly in seed-tissues, where they serve as reserve materials for the early growth of the embryo on germination. The more resistant celluloses and compound celluloses are obviously destined for more permanent functions, and they have been dealt with in this treatise under the term 'permanent tissue.' But permanence is a relative term. We have already endeavoured to trace the changes and modifications which these substances undergo in the normal life of the plant: lignification, for instance, has been dealt with as a progressive chemical transformation of celluloses, differing probably *ab initio* from those of the normal type; and evidence has been given in terms of the constitution of these compounds, showing them to be highly reactive and susceptible of modification in various directions under treatment in the laboratory.

We have now to follow the fate of these substances in the ordinary processes of the natural world. Under these circumstances, 'death' is succeeded by a variety of processes of decay and destruction. These are in part intrinsic, in part determined by outside agencies; they are of the two kinds: processes of resolution, and processes of combination or condensation, which are usually concurrent. The former have been dealt with on p. 66. These are fermentation processes attended with evolution of gaseous products. The latter are defined by an extended series of their products, viz. humus, peat, lignite, and the coals. The chief characteristic of this series of degradation products of plant tissues is the accumulation of carbon at the expense of oxygen and hydrogen. This is illustrated by the following results of elementary analyses, calculated in all cases on the ash-free substance (W. A. Miller, Org. Chem. ed. 1869, 139-146).

	Oak wood	Decayed oak wood	Humus from decayed oak	Peat	
				Dartmoor	Vulcain
C	50·20	53·50	54·0	56·0	59·73	59·57
H	6·08	5·16	5·1	4·9	5·91	5·96
O	43·74	41·34	40·9	39·1	31·82	32·38
N	—	—	—	—	2·54	2·09
Approx. formula (C,H,O)	$C_{24}H_{18}O_{22}$	—	—	—	$C_{20}H_{22}O_8$	—

Coals.

	Lignite Bovey	Scotch	Wigan Cannel	Newcastle	Anthracite
C	67·85	78·46	82·29	87·97	91·87
H	5·76	8·11	5·68	5·31	3·33
O	23·39	13·73	$\{8·31\}$	6·72	$\{3·01\}$
N	0·58		$\{2·18\}$		$\{0·84\}$
Approx. formula (C,H,O)	$C_{27}H_{26}O_7$	$C_{23}H_{32}O_3$	$C_{26}H_{20}O_2$	$C_{45}H_{34}O_2$	$C_{40}H_{16}O$

The coals are chemical aggregates of altogether unknown constitution. By certain observations, however, they are definitely connected with the earlier members of the above series; thus they yield substitution derivatives with chlorine (Cross and Bevan, Chem. News, 44, 185), and yellow acid products on treatment with nitric acid, of similar composition to those obtained from humic compounds, natural and artificial (*infra*). The products of their destructive distillation (coal tar), which it is needless to say have been exhaustively investigated, throw, it is true, a certain light on their constitution; but they are too complex for the establishment of definite relationships to the parent molecules.

HUMIC COMPOUNDS: HUMUS.—These series of compounds

are normal constituents of all soils in which they fulfil important functions. They are produced in the ordinary decay of buried vegetable matter. They are generally of acid properties, and are dissolved by alkalis to brown solutions; they have the characteristics of unsaturated compounds, being readily acted upon by the halogens. They closely resemble the products of decay of forest woods under ordinary conditions. As chief constituents of such mixtures, Mulder recognised geic acid, $C_{20}H_{12}O_7$; humic acid, $C_{20}H_{12}O_6$; and ulmic acid, $C_{20}H_{14}O_6$. These names and formulæ, however, have little more than empirical value. Products closely similar to these are obtained on a large number of decompositions of the carbohydrates, both simple (sugars &c.) and complex (starch, cellulose, &c.). Those obtained by long boiling of the sugars with dilute acids have been carefully investigated by Sestini (Gazzetta, 10, 121, 240, 355). The products are of two classes: (a) soluble in alkalis, sacchulmic acid, $C_{44}H_{40}O_{16}$; (b) insoluble, sacchulmin, $C_{44}H_{38}O_{15}$; both yielding, however, the same substitution products with the halogens, of which the following may be mentioned: dichloroxysacchulmide, $C_{11}H_8Cl_2O_6$ (or $C_{44}H_{32}O_8O_{24}$), and sesquibromoxysacchulmide, $C_{22}H_{18}Br_3O_{11}$ (or $C_{44}H_{36}Br_6O_{22}$). Products very similar to these have been obtained by the authors from the alkaline by-products of the esparto pulping process, viz. by acidification and purifying the precipitate, a body of constant composition, $C_{42}H_{48}O_8$, converted by treatment with HCl and $KClO_3$ into the derivative $C_{44}H_{46}Cl_xO_{20}$ (J. Chem. Soc. 38, 668; 41, 94). The lignic acids described by Lange (see p. 213), and obtained by the action of alkaline solutions upon the woods, are also of similar composition.

On further treatment (fusion) with the alkaline hydrates, these products yield 'aromatic' derivatives, of which protocatechuic acid is obtained in largest quantity (Lange, *loc. cit.*; Demel, Monatsh. 3, 769). It must be admitted, however, that the entire group of products under consideration is extremely ill-defined, and requires to be much more exhaustively investigated to establish definite relationships with the carbohydrates from which they result.

Reverting to the aspect of their natural history, it is clear that the functions of this diversified group are completed in a very remarkable way by this property of carbon condensation. In proportion as they are attacked by destructive agencies, the residue tends to constitute itself into a complex of increasing resistance ; and so the chemistry of the vegetable world, which depends in its proximate relationships upon the properties of polyhydroxy derivatives of the C_6 unit, is ultimately a most striking manifestation of the properties of the carbon atom itself.

PART III

EXPERIMENTAL AND APPLIED

The foregoing general and theoretical treatment of the subject will carry with it a number of suggestions to readers of original and independent habit of mind. Students accustomed to the critical appreciation of essays in the ordering of such subject-matter will have noted the imperfect state of development to which the investigations hitherto prosecuted have brought our knowledge of the celluloses and compound celluloses; and an appreciation of these imperfections implies the possession of the key to the effectual remedy, i.e. to those directions of investigation most certain to lead to valuable contributions to the subject. Students, however, are by no means generally of this class. The habit of initiation and enterprise in scientific work is perhaps rare, not because of want of capacity, so much as of opportunity and the abiding stimulus of necessity. A main purpose of this treatise is to encourage original investigation by opening out, in more definite terms than has been done, the view of a region of the *res publica naturæ* extremely rich in possibilities. This view should not be limited, moreover, to the more ardent few who have imaginative and speculative tendencies. The mastery of method is in itself a useful objective, and the subject of cellulose involves a number of experimental methods which are not merely essential to the work of pioneer investigation, but a necessary part of the general training of the chemist.

In the following pages attention is directed specially to experimental methods, whether for purposes of demonstration or of training in analytical processes of general usefulness. These methods have been described in general terms in the earlier sections; and for the present purpose we shall continue to avoid minute description of practical details, referring the student to the easily accessible accounts of the actual processes to be found in current literature. At the same time, suggestions are included for a course of study, whether of general treatises or special publications on particular subjects, designed to familiarise the student with the work of investigators in this field of natural science.

Laboratory and Research Notes.

Morphology of Cellulose.—The study of cellulose and of the plant fibres generally involves on every hand questions of form and structure, i.e. the form and dimensions of the individual cell or fibre, and the structure or anatomy of the tissue of which it is a unit member.

Microscopic examination and investigation are, therefore, the necessary complement of the chemical study of vegetable substances.

The histological study of minute structure is dealt with in all the standard text-books of the microscope. The structure and elaboration of plant tissues, and the anatomy of the exogenous and endogenous stems are fully described in the text-books of botany which must be consulted. Goodale's Physiological Botany (Ivison: New York) is especially to be recommended.

The most important works treating specially of the minute structure of plant fibres are Wiesner, Mikroscopische Untersuchungen (1872), Die Rohstoffe des Pflanzenreichs (1873); Vétillart, Études sur les fibres végétales textiles (Paris, 1876). Useful information will also be found in Spon's Encyclopedia of the Useful Arts, article 'Vegetable Fibres'; and 'Indian Fibres and Fibrous Substances' (Cross, Bevan, and King: London).

As a useful application of microscopic method the student

should sow some flax (linseed) and examine the stem at intervals throughout the growth of the crop, noting the development of the bast fibre—which constitutes commercial flax—and its structural relationship to wood and cortex.

Mount as permanent preparations the typical paper-making celluloses: cotton, flax, hemp, wood, esparto, and straw. Note dimensions and typical characteristics. Study the photomicrographs of raw materials (sections appendix), And an account of methods of photographing, in Indian Fibres, p. 15.

Bleaching, or Isolation of Cellulose from Raw Fibres.—Take flax and jute as typical raw materials.

Flax.—Extract the fibre in continuous extraction apparatus (Soxhlet) with fat solvents—e.g. alcohol—ether. Make quantitative estimations, and for study of properties of oil-wax extract see Cross and Bevan, J. Chem. Soc. 57, 196.

Boil extracted residue with 2 p.ct. solution NaOH one hour. Examine solution for pectic bodies and note properties. Wash, and bleach fibre in 0·5 p.ct. solution sodium hypochlorite; wash off, treat with sulphurous acid, wash, dry, and weigh. Note that the fibre though colourless has a blackish look. A second portion boil as before, wash, and digest in bromine water some hours. Wash and boil out in carbonate of soda solution. Wash and return to bromine water. Repeat till brilliant white cellulose obtained. Note the gradual disintegration of the woody portions ('sprit') which adhere to the fibre.

Jute fibre.—Boil 15 minutes in 1 p.ct. NaOH solution; wash off, squeeze, and expose to chlorine gas one hour. Note formation of yellow chlorinated derivative of non-cellulose. Wash off and place in 2 p.ct. solution sodium sulphite. Raise to boil, adding a little NaOH. Wash off and treat residual cellulose with sulphurous acid. Wash off, dry, and weigh.

Pine wood.—Take deal shavings or 'wood-wool' and proceed as with jute, repeating the treatment with chlorine until pure cellulose obtained.

Contrast these laboratory methods with those of the bleach works (cotton, linen) and the paper-maker's 'pulping' processes, for which see the standard text-books on these subjects, more especially Chemistry of Paper Making (Griffin & Little: New York, 1894).

Note the correlations of lustre and external appearance with minute structure.

RAPID-COMBUSTION METHOD.—Study method of combustion with chromic acid in presence of sulphuric acid (Cross and Bevan, J. Chem. Soc. 53, 889).

Inorganic or Ash Constituents.—Burn specimens of pure cellulose and raw materials and estimate ash. Note that the ash constitutes an inorganic skeleton of the original. Soak a cotton fabric (muslin) in solutions of ammonium chromate, aluminium acetate, or magnesium acetate; dry and burn. Note that a coherent ash is obtained preserving the structure of the original fabric. A process of this kind is used in making the hoods for the 'incandescent' gas burners; oxides of the rare earth-metals being used for the purpose.

Take an ordinary filter or blotting paper; estimate the ash, and a second portion digest in dilute hydrofluoric acid in a lead or platinum vessel. Wash off with distilled water, dry, and determine the ash. It will be found to be considerably reduced, owing to the removal of silica and siliceous compounds. The ash in cellulosic raw materials is much higher than in the bleached celluloses; notably in the straws. The proportion of silica is exceptionally high. It was frequently affirmed in the older text-books that the rigidity of the straws was due to this constituent, but special cultivations in absence of silica give a straw of normal appearance and rigidity. The structural properties of these stems are therefore referable to their 'organic' components.

Hydration and Dehydration.—For a wider discussion of hydration and dehydration types as applied to the elucidation of the special chemistry of the carbohydrates, the student should read a paper by Baeyer, 'Ueber die Wasserentziehung und ihre Bedeutung für das Pflanzenleben u. die Gährung' (Berl. Ber. 1870, 363); or the translation by Armstrong (J. Chem. Soc. [2], 9, 331).

The *absorption, osmotic and capillary phenomena* resulting from the interaction of cellulose and aqueous solutions should be carefully studied. Much work remains to be done on this subject. The student should acquaint himself with Pfeffer's methods of determining osmotic pressures, and the bearings of the phenomena of osmosis on the theory of solution.

Cellulose Solvents.—Prepare solutions of cotton and other celluloses by treatment with zinc chloride. To dissolve 1 part cellulose take 4-5 parts $Zn.Cl_2$, dissolve in 5-7 parts water, add the cellulose, heating in a porcelain dish over a water-bath. Stir from time to time, and add water to replace that which evaporates.

Precipitate the solution (1) by pouring into water. Wash till free from Zn salt, dry, and weigh. Burn off the cellulose, and estimate the residue of $Zn.O$.

(2) Pour into dilute hydrochloric acid. The precipitate will be cellulose hydrate free from ZnO. Wash thoroughly, dry, and weigh.

Calculate to original cellulose, and show that the molecule is hydrated. Control by examining the solution in which the cellulose was precipitated for dissolved products of hydrolysis.

(3) Precipitate by pouring the solution into alcohol in a fine stream, or by spreading the viscous solution as a thin film on glass. Submerge the whole in alcohol, and detach the coherent film of the cellulose compound. Digest with HCl.Aq to remove $Zn.O$. Wash, and dehydrate by lengthened exposure to alcohol.

The progress of the action of the zinc chloride solution should be followed up microscopically. Cotton fibres are mounted in the strong solution, covered with cover glasses, and the glass slips ranged on a hot surface, so that they may acquire a temperature of 80-90°. Examine from time to time under low and high power, noting the progress of the disintegration, swelling-up, and rupture of the cell wall. The structural points may be further differentiated by staining with iodine. To apply the iodine without precipitating the dissolved cellulose it is necessary, after dissolving the iodine in a little potassium iodide in the usual way, to dilute with 40 p.ct. zinc chloride solution. A drop of this solution may be added to the solution on the slide without otherwise affecting the cellulose than staining the fibrous residues.

Dilute ammoniacal solutions of cuprous oxide are without action upon cellulose. An interesting demonstration of the difference between the cuprous and cupric oxides in this respect may be carried out as follows : Cuprous chloride is prepared by the action of hydrochloric acid upon copper, with addition of potassium chlorate in successive small quantities. The chloride is washed with water free from oxygen, and, after settling, portions of the magma of crystals and water are placed upon a filter paper, which is then transferred to a bottle provided with an indiarubber cork. Through the cork, glass tubes are passed, and so arranged that a stream of coal gas may be led into the vessel to completely expel the air. After expelling the air, a little strong ammonia solution is introduced by means of a separating funnel previously inserted through the cork. No change in the substance of the paper is observed. The stream of coal gas is now stopped, and atmospheric air is introduced. Reaction rapidly ensues, with development of the blue colour of the ammoniacal cupric compound, accompanied by solution of the cellulose. The details of the arrangement of the experiment are immaterial, and may be varied in a number of ways.

On the theory of the action of cellulose solvents, read a paper by the authors in Bull. Soc. Chim. 1893, 295.

Cellulose Xanthate.—In preparing the solution of the cellulose thiocarbonate for use in the laboratory it is better to employ a 'rag' cellulose (cotton and linen) disintegrated by the process of 'beating' in a paper mill, which can be easily procured. If obtained in the moist condition the cellulose may be air-dried previously to treating with the 15 p.ct. solution of NaOH. The cellulose and the alkaline lye may be thoroughly incorporated by grinding together in a mortar in the calculated quantities to finish the alkali cellulose so as to contain

Cellulose, 100; caustic soda (NaOH), 45; and water, 250-300;

or the cellulose may be treated with the 15 p.ct. NaOH solution in excess, and, after standing some time, the cellulose may be drained off on a filter of perforated zinc and squeezed to retain three times its weight of the solution.

As the process of 'mercerisation' requires some time for completion, it will be found advantageous to set aside the moist alkali cellulose in a closed bottle for two or three days before subjecting it to the solvent reaction. This reaction proceeds spontaneously at the ordinary temperature; the alkali-cellulose and carbon disulphide (40-100 parts cellulose) are brought together in a stoppered bottle, vigorously shaken together for a minute or two, and set aside for two or three hours. The resulting yellowish mass is covered with water and allowed to stand some hours, then vigorously stirred with addition of water, in quantity calculated to produce a solution of any desired percentage strength (cellulose).

The progress of the reaction should be followed by microscopic examination of the fibre at intervals. The rupture of the cell walls is preceded by an exaggeration of the structural details of the fibre, which will be found useful in differentiating cotton from celluloses of other types.

From the solution of the crude thiocarbonate the solid product is precipitated, in the form of a gelatinous hydrate, by various 'neutral' dehydrating liquids or solutions: of these the most convenient are alcohol and a solution of common salt, and the *modus operandi* may be instructively varied in the following ways:—

(1) *Alcohol.*—The solution may be poured into a photographer's developing dish to a depth of, say, $\frac{1}{8}$ inch, strong alcohol may then be poured upon the solution, the dish covered with a glass plate, and set aside. Coagulation of the cellulose product proceeds gradually, and after some time a coherent slab is obtained of a greenish colour, the yellow by-products of the reaction having been dissolved by the alcohol.

Or the solution may be caused to flow in a fine stream into the alcohol, when the cellulose xanthate (hydrated) will be precipitated as a continuous gelatinous thread. In the latter case it is advisable to add a certain quantity of alcohol to the crude solution, which can be done without causing precipitation, the xanthate being soluble in dilute alcohol. Such a solution will obviously be more sensitive to the further action of the alcohol, and precipitation in slender and uniform threads is facilitated.

(2) *Brine.*—If prepared from the commercial salt, the solution should be freed from magnesia by previous treatment with sodium carbonate and filtering. The precipitation may be carried out as described under (1), or any surface may be coated with the crude solution, and the xanthate precipitated upon the surface by immersion in the brine. The xanthate, purified by any of these treatments, may be redissolved in water, and again precipitated by a repetition of the treatment. With each precipitation the ratio of both alkali and sulphur to cellulose in the product is considerably diminished.

The precipitated xanthate may be treated with solutions of

suitable salts of the heavy metals, and the cellulose xanthate of the metals prepared.

Cellulose is regenerated from the solution in various ways which may be practised by the student : (*a*) the solution may be set aside in any suitable vessel for some days ; (*b*) it may be heated at 80-90° in a water bath ; (*c*) it may be spread upon a glass plate or other surface, dried at 60° C., and finally heated for some minutes at 100° C.

The masses or films obtained are washed to remove the alkaline by-products ; and the cellulose bleached, if necessary, by treatment with sodium hypochlorite in dilute solution (0·5–1·0 p.ct. NaOCl). Of the special uses of the solution in the laboratory, the following may be particularised :—

(1) *In Microscopic Work.*—In preparing fibres for cutting in cross section, they may be worked up with the viscous solution into a strand or pencil, the cellulose being regenerated spontaneously or otherwise. Such a strand may be cut with or without a microtome, or may be further 'embedded' in cellulose by submerging it in the viscous solution and coagulating. In some cases the crude solution may be used ; in others the solution should be previously discolourised (and neutralised) by treatment with sulphurous acid.

(2) In diffusion experiments (osmosis) the regenerated cellulose is of use in the preparation of membranes or diaphragms, the cellulose being either used alone, or a compound diaphragm may be made by coating paper or cloth with the thick solution, and 'fixing' the cellulose by any of the methods described.

Important results may be expected from a study of osmotic transmission through this form of cellulose. The passage of crystalloids through the cellulose, especially in its hydrated forms, appears to be exceptionally rapid.

Theoretical Notes.—Determinations have been made in which known weights of pure cotton cellulose have been dissolved as thiocarbonate, the cellulose regenerated, spontaneously and in other ways, then carefully purified and weighed. The cellulose recovered shows a slight increase upon the original. It may be taken, therefore, that the molecule undergoes no permanent disaggregation under the treatment, and it is even probable that the cellulose maintains a high molecular weight in solution. In contrast to its behaviour under the severe alkaline treatment, it has been noted that when dissolved in zinc chloride, more especially in presence of hydrochloric acid, there is a progressive hydrolysis of the molecule, with conversion into soluble products.

It is evident, therefore, that the *hydration* of cellulose has nothing in common with *hydrolysis*; the cellulose molecule appears to have an indefinite capacity for combining with water and for undergoing an extensive series of hydration changes without affecting its fundamental constitution.

Cellulose is usually regarded as a 'condensed' derivative, presenting more or less analogy with starch. But in resistance to hydrolysis there is a marked and very important distinction, the due interpretation of which will no doubt result in establishing for the celluloses a special constitutional type.

We write the cellulose unit as $C_6H_{10}O_5$, remembering that this merely represents the empirical facts of its ultimate elementary composition. It is of course evident that at least one of the O atoms is present as carbonyl oxygen. Further, we may regard the constituent groups of the molecule as grouped around the CO or negative, and a CH_2 or positive centre. There are many evidences of a 'polarity' of this character manifested by cellulose, as by other 'carbohydrates,' in reaction. The thiocarbonate reaction proper we may regard as localised at the more negative OH groups, i.e. those in proximity to CO, which are held in combination by the alkali used in mercerisation or the formation of alkali cellulose.

The attendant phenomena of mercerisation indicate a considerable degree of hydration of the cellulose (combination with water), associated with the entrance of the alkaline groups. This gelatinous condition of the alkali cellulose hydrate conforms with the characteristics of the earlier stages of the process of solution of colloids. The completion of the process requires in this case the additional strain of the carbon disulphide, evidently acting as an acid group. It is certainly noteworthy, as already pointed out, that all the solvents of cellulose (as cellulose) are of a saline character, and may be regarded as forming with the cellulose, by reciprocal combination with its acid and basic groups, compounds analogous to the double salts. The actual mechanism of solution we cannot pretend to follow. The modern theory of solution in the hands of its most advanced exponents confines itself to the investigation of the molecular condition of substances *in* solution; and, in the case of electrolytes, it is sufficiently established that solution is attended by dissociation into ions.

This aspect of solution may very well be taken up in dealing with the solution phenomena of the carbohydrates. The lower members of the series, in solution, show in most cases considerable sensitiveness to the action of acids and bases, and form a number of 'molecular' compounds with salts such as sodium and calcium chlorides—which may very well be considered as double salts. This indicates a saline character of the molecule in solution. The phenomena of alcoholic fermentation also point in the same direction. The resolution of a dextrose molecule into alcohol and carbonic acid has entirely the characteristics of an electrolytic split; and it is fair to assume that the molecule in aqueous solution already manifests a 'strain' in the direction in which cleavage ultimately takes place. In regard to cellulose in solution, it is also reasonable to assume a similar 'polarisation'; and the reactivity of the cellulose regenerated from solution is explained by assuming that this polarity is in a measure retained.

In regard to the hydrates of cellulose, and the extraordinary power which cellulose has of solidifying water, it appears that cellulose must have the capacity of absorbing into itself the 'energy of condition' of water.

It would be important to determine the physical constants —volume and vapour tension—of the water, in the form of a series of definite hydrates of cellulose. This would afford a measure of transference of energy from the water to the cellulose.

Cellulose Benzoates.—These reactions have been studied only under ordinary conditions of temperature, and under conditions where there is considerable loss of the anhydrochloride in waste reaction with the alkali. This impedes the direct synthetical reaction between the cellulose and benzoyl residues, and causes large variations of the product. The reaction is necessarily difficult to control, especially in the case of the insoluble alkali celluloses, and requires further systematic investigation from this point of view, which will also result in the formation of definite and uniform products, i.e. so far as is attainable with a reaction in which the original substance and derivative are both insoluble in the reaction medium.

The conditions requiring systematic variation are (1) the

temperature: the reaction should proceed at the lowest limit of temperature; (2) the conditions of mercerisation: the reaction would probably be favoured by employing a certain proportion of zinc oxide in replacement of sodium hydrate; and (3) the liquid medium. There is, of course, a rapid hydrolysis of the benzoyl chloride by the solution of sodium hydrate; and there are various ways which suggest themselves of retarding this waste reaction.

The precipitation, as benzoates, of modifications of cellulose soluble in alkaline solutions should prove a valuable aid to investigation; but, before the method can be applied to the investigation of substances and mixtures of unknown or lesser known composition, it will be necessary to characterise the products obtainable from the normal celluloses. There are in plant tissues a widely diffused group of substances having the external characteristics of the resistant celluloses, or celluloses proper, but which are readily soluble in alkaline solutions. In the course of an ordinary proximate analysis these are usually 'dismissed' as 'non-nitrogenous extractive matters'; which, of course, is but a very loose description.

We have seen that hydrated modifications of the normal cellulose are themselves soluble in alkaline solution, and it would appear that in this condition they are generally more reactive.

There are two practical problems suggested by the properties of these hydrates: (1) Do they yield to the processes of animal digestion? and (2) how far can the processes adopted by the paper-maker for isolating cellulose from complex fibrous materials be limited, to prevent the solution of celluloses either existing or hydrated in the raw material or tending to pass into such modifications?

The investigation of these questions will be found to be greatly facilitated by a method of separating the soluble celluloses from solutions. The benzoates give promise of affording such a method of separation, and invite investigation from this point of view as much as from that of their intrinsic theoretical interest.

Cellulose Acetates.—The 'suppression' of the OH groups of cellulose is evidenced by its not reacting with acetic anhydride at its boiling point; whereas the poly-hydroxy-compounds generally are acetylated under these conditions, and many even react with acetic acid itself. We may find some explanation of this in the evidence that we have of a species of saline or ethereal constitution which characterises cellulose in many of its reactions. It is probable that the OH groups of opposite function exert a mutually repressive influence; and acetic anhydride, being essentially a condensing agent, is unable to determine the liberation of the OH groups into the condition in which they can react.

On the other hand, the presence of certain reagents in relatively small proportion is sufficient to disturb the equilibrium, and reaction results. An elevated temperature also determines reaction, and obviously for a similar reason, viz. that the equilibrium of the molecule holds only for a certain range of conditions.

For these reasons it may be considered doubtful whether a cellulose acetate in the strictest sense can be obtained. Acetates are, however, obtainable which satisfy the 'general' *à priori* definition—viz. which by saponification yield a carbohydrate having the negative characteristics of cellulose, and in contradistinction to acetates which, though obtained from cellulose, yield on saponification a carbohydrate more or less soluble in the alkaline solution, and reducing CuO.

Parchmentising Process.—The student should make careful statistical observations upon this reaction. An unsized paper should be used—ordinary 'waterleaf' or filtering paper. The acid should be placed in a 'photographic' dish of suitable size; the sheet immersed in this and, after suitable exposure to the action of the acid, transferred at once to a dish of water. After staying in this first wash water for some time they must be thoroughly washed in a flow of water until neutral. The sheets having been weighed before treatment, and the moisture estimated in portion of the same paper—the product must also be weighed after drying. It is necessary also to determine the cellulose dissolved by the process, for which purpose the acid and first wash water may be treated. These may be mixed and a certain fraction boiled, the dextrose formed being estimated by any of the well-known methods; or the solution may be neutralised with lime evaporated, and the total carbon estimated, by the method of combustion, with CrO_3 and H_2SO_4.

The product may also be further examined (1) for loss of weight in boiling with dilute alkaline solutions; (2) CuO reduction in presence of alkalis (Fehling's solution); and (3) degree of acetylation, which is a measure of the OH groups rendered 'free' or reactive by the process. A systematic series of observations embracing these points would be a useful contribution to our knowledge.

'Toughening' Action of Nitric Acid upon Papers.—The student should read the original paper of Francis (J. Chem. Soc. 47, 183) and repeat his observations on the toughening of filter paper by this process.

Preparation of Hydrocellulose.—This reaction may also be studied by the student statistically. A given weight of a pure bleached cellulose fabric, preferably in the form of rag, is exposed, in the air-dry condition, to hydrochloric acid gas in a closed vessel. The reaction is complete when the fabric breaks down on slight pressure to an impalpable powder. It may, however, be more conveniently washed in the original form of cloth, which it will retain sufficiently—notwithstanding the disintegration which has taken place—to withstand a current of wash water, carefully admitted to the bottom of the vessel. The product when pure may be dried and weighed, and its properties and reactions determined.

The student should convert the product into the corresponding nitrates. One part of the product (dried at 105° C. and cooled in a dry atmosphere) is gradually stirred into 5 parts of concentrated nitric acid (HNO_3 of 1·5 sp.gr. and colourless). The gummy solution of the nitrate may be treated in various ways for the separation of various products.

(*a*) It may be diluted with a small proportion of water, gradually stirred in so as to keep the product dissolved; the solution then poured on to a glass plate, and the acid evaporated at a temperature of 40° C. in a free draught of air. The product is then washed, and detached as a film from the glass.

(*b*) The concentrated or somewhat diluted solution may be poured into water or spread on a glass plate, which is then plunged beneath water contained in a suitable vessel.

(*c*) The precipitation may be effected by sulphuric acid—the acid being added drop by drop to the concentrated solution, or the solution poured into the sulphuric acid. In either case the mixture must be carried out without sensible rise of temperature.

Preparation and Diagnosis of Oxycelluloses.—The student of cellulose technology will find it necessary to study the original papers (*loc. cit.*), and to repeat the experimental demonstrations therein given, of the formation of these oxycelluloses, and the methods of diagnosing their presence in textile fabrics.

It is to be noted that the ordinary bleached celluloses of the cotton group all give a small proportion of furfural (0·3-1·0 p.ct.) on boiling with hydrochloric acid (1·06 sp.gr.), which may be taken as indicating the presence of an oxycellulose in proportionate quantity. It is probable also that the somewhat increased reactivity of bleached cotton may be due to formation of oxycellulose.

Study of Methods of Bleaching.—The student should compare the bleaching actions of potassium permanganate and sodium hypochlorite—from the point of view of bleaching effect, and relative consumption of oxygen. It must be noted that the permanganate (i.e. Mn_2O_7) is deoxidised to the dioxide MnO_2, which is deposited as the brown-coloured hydrate upon the cellulose; the depth of colour is therefore a measure of the oxidation which has taken place, generally or locally. The oxide is removed by treating the substance, after washing from the alkali simultaneously set free, with a solution of sulphurous acid, the interaction of the reagents producing dithionic acid. With the disappearance of the brown oxide the bleaching effect produced by the original oxidation is at once apparent.

While the action of the permanganates upon the cellulose is necessarily that of simple oxidation, the hypochlorites may act in two ways: (1) a simple oxidation; (2) chlorination. The latter effect is usually small, the proportion reacting in this way depending upon the temperature of the solution and also the nature of the base with which the hypochlorous acid is combined. See 'Some Considerations on the Chemistry of Hypochlorite Bleaching' (J. Soc. Chem. Ind. 1890).

Formation of Acetic Acid from Cellulose.—The student should read the account of a systematic investigation of these reactions by J. F. V. Isaac and the authors (J. Soc. Chem. Ind. 1892).

The maximum yields of 30–40 p.ct., which are obtained under the most favourable conditions, point to a $CO-CH_3$ 'residue' in the cellulose molecule itself, as the immediate source of the acetic acid.

Ferment-hydrolyses of Cellulose.—It is obvious that chemical compounds destined to transmit, accumulate, and otherwise respond to, the form of energy which we are bound to express by the somewhat colourless term 'vital force' must exhibit a specially plastic constitution. The peculiar sensitiveness of the carbohydrates as a group to fermentation and other decompositions is not sufficiently accounted for by the constitutional formulæ assigned to them.

Destructive Distillation.—*Theoretical Notes.*—We may by careful consideration form some mental picture of the consequences of the addition of heat to compounds of this class. As a preliminary it is necessary to remember generally the distinction between the purely chemical and the chemico-physical view of the constitution of matter. According to the former, cellulose is a compound $n.C_6H_{10}O_5$, of which the ultimate constituent groups—CO—CH$_2$—CHOH—are arranged in a certain way in the C_6 units ; and these, again, are grouped together in a special configuration which, when elucidated, will represent the molecular constitution of cellulose. The correlative of this view is that of the intrinsic or internal energy of the compound. This is expressed as a crude aggregate —on the thermo-chemical view—as what is known as the 'heat of formation,' which, in the case of carbon compounds, is the difference between the sum of the heats of combustion of its constituents, and the heat of combustion of the compound itself, expressed in arbitrary units of mass, usually the gramme, but in molecular ratios. The methods of determining combustion-heats are necessarily subject to errors of observation of some magnitude ; but, even with much closer approximation to accuracy, the number of calories per gram-molecule of a compound evolved on burning is still of only empirical and statistical value, even when applied to compounds related in differential series.

Whatever the meaning of the constant of energy, which we regard as associated with every particular configuration of matter, it is at least obvious that all compounds with which we are familiar represent an equilibrium of matter and energy under the conditions of observation ; and as combination is generally associated with loss of energy, so, *vice versâ*, the introduction of energy implies decomposition or tendencies to decomposition. The communication of energy in the form of the electric current, as in electrolysis, is usually attended by more or less simple decompositions, more or less easy, therefore, to follow. The effects of heat, on the other hand, are complicated by recombinations of the products amongst themselves, more especially in the regions of high temperatures, rendering it somewhat difficult to distinguish the primary from secondary products.

Generally we may take the saturated compounds of lower molecular weight as direct products of the decomposition: these are water, oxides of carbon, methane, methyl alcohol, formic and acetic acids, &c. Volatile compounds, either hydrocarbons or their derivatives, containing unsaturated C—H nuclei, result from rearrangements of the component groups of the original celluloses; these are olefines and aromatic hydrocarbons, furfural, phenols, aromatic methoxy-derivatives, &c.; the residue of the process—or charcoal—represents the extreme limits of condensation of the carbon nuclei.

One important general feature of the decomposition is noticeable as connecting the decomposition with those determined by electrolysis and fermentation in the case of compounds of similar constitution, that is, the accumulation of hydrogen in the one direction, and oxygen in the other. Thus we have, on the one hand, methane and a large number of compounds containing the CH_3 and $O.CH_3$ group; and, on the other, CO and CO_2. Acetic acid represents an intermediate equilibrium, viz. of CH_4 and CO_2. We have already seen that the introduction of the alkaline hydrates so alters the character of the decomposition that acetic acid becomes a main product of decomposition; and the remainder of the reaction consists in the production of the more fully oxidised oxalic acid (with CO_2), 'balanced,' so to speak, by the formation and liberation of hydrogen. These considerations illustrate the views which may be applied to the elucidation of the very complex problems in dissociation, generally presented by destructive distillation.

Constitution of Cellulose.—*Theoretical Notes.*—It is, perhaps, premature to approach the problem of the constitution of cellulose, as the molecule of cellulose is at present an altogether unknown quantity. It will be instructive to the student to study, for their general bearing on the subject, the investigations which have been devoted to the elucidation of the structure of the starch molecule, notably those of O'Sullivan (J. Chem. Soc. 1872), Brown and Morris (*ibid.* 1889, 449-462), Lintner and Düll (Berl. Ber. 26, 2533).

These researches are based on the study of the hydrolytic dis-

S

section of the molecule, which breaks down ultimately into maltose when invertase is employed, or into dextrose when the hydrolysis is completed by acids, the intermediate terms being the uncrystallisable dextrines. Owing to the obviously simple relation of the products of resolution to the original molecule, we are justified in concluding that starch represents an aggregate of considerable molecular magnitude; the ultimate constituent C_6 groups being linked together by oxygen. The evidence, in fact, goes to show that starch is of the same constitutional type as the bioses (e.g. cane sugar and maltose) and trioses (e.g. melitriose). The combination of two glucose residues, to form a biose, takes place in one of two ways— i.e. either one or both of the typical CO groups takes part in the condensation; in the former case (e.g. maltose) the free carbonyl determines aldehydic properties, which disappear, in the latter case, (e.g. cane sugar) with the suppression of the second CO group.

In starch, the linking of the glucose residue is of the dicarbonyl type, and it consequently fails to react with phenylhydrazine, and does not reduce Fehling's solution. In the hydrolysis of starch the carbonyl linkings are broken down successively, the process being similar to that which takes place with the triose above named, which is first resolved into melitriose and fructose, the former then splitting into glucose and galactose. (See Scheibler and Mittelmeier, Berl. Ber. 26, 2930.)

Cellulose is identical with starch in empirical composition ($n.C_6H_{10}O_5$), and similar, in being an aggregate of hexose groups; but for resolution into the latter, cellulose requires the severe treatment of solution in concentrated sulphuric acid.

To give full value to these characteristic differences, we must pronounce for a corresponding difference in molecular configuration.

The problem must next be considered from the point of view of function of the reactive groups of cellulose. The student should be reminded that carbon chemistry is very much a chemistry of *function*. The idea of chemical function grew but slowly with the study of the 'inorganic' elements, and indeed of the carbon compounds also, so long as it was limited to individuals and groups more or less isolated. It was a chemistry of jumps and gaps. In modern 'organic' chemistry, on the other hand, the carbon compounds are presented to us in extended series of progressive differentiations—which are therefore, in the limited sense of the term, 'organically' related. The differentiations are of two kinds : (1) the reactive groups (CO, OH, &c.) may vary in number and position ; (2) the substantive group (C_nH_m) may vary in configuration (aliphatic, cyclic, condensed, &c., hydrocarbons). To trace the corresponding variations in function or reactivity is a very important part of every investigation ; and the student should diligently study the papers of original workers, and familiarise himself with such general expositions of the science, from this point of view, as he will find in the opening chapters of Beilstein's great work, Handbuch der organischen Chemie ; and more especially, in regard to the carbohydrates, in Tollens' monograph, Die Kohlenhydrate (Breslau, 1888).

The student should endeavour to form a critical estimate of the position of cellulose in relation to starch and others of the 'saccharo-colloids' (Tollens), to which it is most nearly allied. Such an estimate must be based upon the comparative study of reactions.

Comparing cellulose with starch we find the former resistant to hydrolysis and acetylation, but giving the highly characteristic thiocarbonate reaction. These differences are differences of function or reactivity of OH groups on the one hand, and of the linking together of the unit groups on the other.

We have then to consider whether these differences are sufficient to constitute the cellulose group a special type of constitution. We think they are.

Assuming the unit group $C_6H_{10}O_5$, and decomposing this, on the evidence before us, into $C_6H_6O.(OH)_4$, we have but few alternatives. The conclusion we draw can only be expressed within the limits of these alternatives, and therefore in general terms, as follows : (1) The C atoms of the unit groups are combined in a closed ring ; (2) the linking of the unit groups is not an oxygen, but a carbon linking ; (3) the synthesis of the unit groups together may be assumed to occur between the CO of one group and the CH_2 of another, giving the alternative form $CH-C(OH)$.

LIGNOCELLULOSES OF CEREALS.—The student should read the special papers on the subject, Berl. Ber. 1894, and J. Chem. Soc. 1894.

The cereal straws—in which may be included the envelopes of the seeds of these and allied species—are of complex constitution. The tissues, in addition to being lignified more or less (see Lignocelluloses, p. 92), contain, as an essential constituent, the characteristic compound known as wood gum (*Holzgummi*). Wood gum is the colloid anhydride of the pentaglucoses, yielding the crystallisable pentoses, xylose and arabinose, when hydrolysed by boiling dilute acids. The pentosans are dissolved by dilute solutions of the alkaline hydrates in the cold, from which solutions they are precipitated on acidification, and more completely in presence of alcohol. The pentosans, when boiled with hydrochloric acid (1·06 sp.gr.), of course yield furfural, and it is sometimes assumed that the quantity obtained from straw is a direct measure of the pentosans present. This conclusion requires qualification, and it is necessary to divide the furfural-yielding constituents into the two groups: (*a*) the pentosans, easily hydrolysed, and obtained as above described; (*b*) the oxycelluloses, resisting hydrolytic actions of some intensity.

Together with the investigations above cited, the student should read an account of the researches of Schulze and Tollens (Landw. Vers.-Stat. 40, 367) upon the composition and constitution of the substance of 'brewers' grains.' This material consists obviously of the more resistant seed-envelopes of the barley, the greater portion of which remains unaffected by the malting and mashing processes. The following scheme represents the method of examination pursued by the authors with indications of the results obtained:—

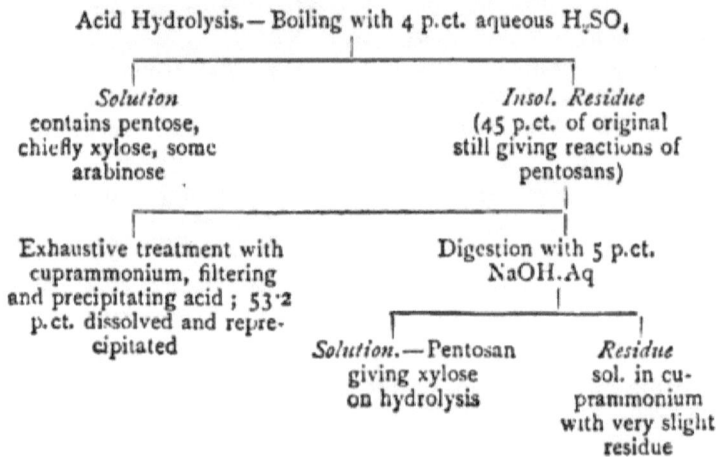

Acid Hydrolysis.—Boiling with 4 p.ct. aqueous H_2SO_4

Solution contains pentose, chiefly xylose, some arabinose

Insol. Residue (45 p.ct. of original still giving reactions of pentosans)

Exhaustive treatment with cuprammonium, filtering and precipitating acid; 53·2 p.ct. dissolved and reprecipitated

Digestion with 5 p.ct. NaOH.Aq

Solution.—Pentosan giving xylose on hydrolysis

Residue sol. in cuprammonium with very slight residue

An important conclusion from these experiments is, that the cellulose and pentosan constituents of these tissues exist together, in combination rather than in mere admixture, the complex also containing the characteristic lignin groups. (See Lignocelluloses, p. 156.)

General Methods for Identification of Carbohydrate Groups
(Hexoses, Pentoses, &c.)

The researches of Tollens and others (Landw. Vers.-Stat. 39, 401) have contributed considerably towards the completion of methods of proximate resolution of such mixtures as it has been customary to define by the term 'non-nitrogenous extractive matters.' In the investigation, more particularly of vegetable food-stuffs and fodder plants, the so-called 'Weende' method of Henneberg has been generally adopted. This consists in the direct determination of fat, protein, 'crude fibre,' and ash; and the sum of these constituents subtracted from 100 is represented as 'non-nitrogenous extractives.' These are the less resistant carbohydrates in question, and concerning these it is important to determine whether they contain (1) the true (hexa) carbohydrates or their anhydrides; and (2) the presence or absence of (*a*) dextrose, (*b*) galactose, (*c*) levulose, (*d*) other carbohydrates, more especially mannose; and (3) the presence or absence of the pentaglucoses. It is also often necessary to estimate these compounds quantitatively. It is also necessary to keep in view the probable presence of derivatives of these compounds or group, more particularly the oxidised derivatives. It will be obvious that no generally applicable scheme of analysis can be laid down; but the following methods of diagnosis are typical, and should be carefully worked out by the student.

(1) *General identification of true (hexa) carbohydrates by levulinic acid reaction* (Tollens, Ann. 243, 315). The substance is heated for 20 hours at 95–98° with HCl.Aq (1·10). Levulinic acid is extracted by exhaustion with ether, converted into the zinc salt (cryst.), and this into the silver salt $C_5H_7O_3Ag$, yielding 48·43 p.ct. Ag on ignition.

(2) *Identification of dextrose groups by formation of saccharic acid.*—The substance is oxidised by digestion with HNO_3 (1·15 sp.gr.); the saccharic acid converted into the acid potassium salt, and this into the silver salt, giving 50·94 p.ct. Ag on ignition.

(3) *Identification of galactose groups by conversion into mucic acid.*[1]—The substance is oxidised with HNO_3 (1·15 sp.gr.); the mucic acid crystallising out directly, owing to its insolubility. Galactose yields 75 p.ct. of its weight of the acid.

(4) *Identification of levulose groups by reaction with resorcinol.*—Levulose may be sometimes identified by the relative ease with which it is converted into levulinic acid. Oxidising methods are not available owing to the ease with which it breaks down into acids of lower molecular weight. The reaction with resorcinol in presence of HCl—a fiery red colouration—may be relied on. The reagent is prepared by dissolving 0·5 grm. resorcinol in 60 c.c. HCl of 1·09 sp.gr., with which the solution to be tested is gently warmed.

(5) *Identification of mannose as phenylhydrazone.*—Mannose is easily distinguished by its property of reacting in neutral dilute solution with phenylhydrazine (acetate) to form an insoluble hydrazone, which may be further identified by determining its melting point (188°).

(6) *Identification of pentaglucoses and oxycellulose.*—These are the furfural-yielding carbohydrates, and their identification and estimation depend upon their conversion into furfural by boiling with HCl (1·06). In quantitative determinations the latter is estimated as hydrazone.

Lignocelluloses.—*Laboratory Notes.*—It will have been evident to the student that the chemistry of the lignocelluloses is that of a highly reactive molecule, and therefore is different in a great many respects from the celluloses. The reactions of the fibre-substance have also been dwelt upon more in detail than in the case of cotton cellulose, and it will therefore be unnecessary to do more than collect at this point those reactions which are typical and characteristic, and which are useful in the laboratory either for demonstration or in the investigation of unknown materials.

(1) **Qualitative Reactions.**—Colour reactions with phloroglucol (solution in HCl), aniline sulphate (aqueous solution) with iodine (absorption with development of brown colour), ferric ferricyanide (deep blue dye) with magenta sulphurous acid, and with

---

[1] The methyl hexoses are similarly oxidised to mucic acid. (E. Fischer, Berl. Ber. 1894, 385.)

coal-tar dyes in great variety (simply dyeing or staining phenomena).

(2) **Solutions of the Lignocelluloses.**—In cuprammonium, in zinc chloride (saturated aqueous solution), and zinc chloride in HCl; *partial solution* under the thiocarbonate reaction. This reaction should be carefully followed up with the microscope, more especially the extraordinary combination which takes place with water on covering the fibre, after the reaction, with water.

(3) **Hydrolytic Agents.**—(*a*) Concentrated solutions of caustic soda (10 25 p.ct. NaOH) in the cold. This reaction should be studied quantitatively, and should be also followed up with a microscopic observation of the fibre under action.

(*b*) *Dilute alkali solutions.*—These constants, quantitatively determined in terms of loss of weight, are important. The usual conditions are a 1 p.ct. solution of caustic soda in large excess; the specimen being boiled for 10 minutes, and a second for 60 minutes, keeping the volume of the solution constant. The fibre is then washed off and treated with a little dilute hydrochloric acid, again washed, and dried.

(*c*) *Dilute acids.*—The observation of the action of boiling dilute sulphuric acid (1 p.ct. H_2SO_4) is of use in differentiating one lignocellulose from another. The fibre may be boiled for 60 minutes with an excess of 1 p.ct. acid, keeping the volume of the solution constant. The fibre is washed free from acid, dried, and weighed.

(4) **Cellulose Estimations.**—On this subject see Cross and Bevan, J. Chem. Soc. 1882. The following methods should be worked:

(*a*) *Bromine method.*—The fibre is boiled in dilute alkali, washed, and placed in saturated bromine water and left for some hours, afterwards washed and boiled in dilute ammonia. The fibre is then washed and returned to the bromine water, and again boiled, after digestion, in ammonia. The treatment is repeated so long as any residues of a yellow colour are seen in the boiling alkaline solution. The cellulose is finally washed with dilute acid and then with water, dried, and weighed.

(*b*) *Chlorine method.*—This involves not only a determination of cellulose, but of the amount of chlorine disappearing in reaction, and also the amount of hydrochloric acid (Read, Cross, and

Bevan, J. Chem. Soc. 1889, 199). Instead of the bulb there described as blown on the end of a tube, a simpler method consists in blowing a thin bulb and inserting the prepared fibre, and after insertion, drawing of the bulb. It is then placed in the reaction flask, containing an atmosphere of chlorine. The flask or bottle is closed with a cork well covered with paraffin, and bored with one hole to admit the glass tube connecting with the measuring apparatus. When the levels are all adjusted, the bulb is broken by a blow against the side of the bottle. For further particulars see the paper above cited, also for the estimation of the hydrochloric acid formed. The cellulose estimation, by boiling the chlorinated fibre with sulphite of soda solution, &c., and washing off, has been already described (see p. 95).

(c) *Nitric acid process.*—The weighed fibre is placed in a flask, and digested with 5 p.ct. nitric acid at 60°. The gaseous products may be collected and examined. Digestion is continued until the yellow colour which at first results gives place to white. If arrested at an earlier stage the residues of non-cellulose may be removed by treating with weak alkaline solution. The cellulose is washed, treated with dilute acid, again washed, and dried for weighing. The yield is from 60–66°, considerably less, therefore, than by either of the above methods.

In reference to the theory of the action, show that by adding urea the specific action of the acid is entirely arrested, and it becomes similar to that of hydrochloric acid or sulphuric acid dilute. The specific action also has a limit in reference to the concentration of the acid, which must contain at least 3 p.ct. HNO_3.

(d) *Chromic acid process.*—This process should be investigated with chromic acid only, and with the addition of acetic acid and of sulphuric acid. In the first case, the action is extremely slow, and there is considerable combination of the oxide with the fibre-substance. The presence of the hydrolysing acid causes the specific oxidation of the non-cellulose constituents. In dilute solution this takes place without evolution of gas. The product may be tested from time to time by washing off a small portion, exposing to chlorine, and afterwards plunging into sodium sulphite solution. When this reaction ceases, arrest the experiment, wash off the product, dry, and weigh. After weighing, compare this oxy-cellulose mixture with the celluloses obtained by the above processes. In experiments where sulphuric acid is added as the hydrolysing acid, the solution should be distilled and the volatile acid estimated.

(*e*) *Alkali oxidations.*—Study the continued action of hypochlorite of sodium and of hypobromite, continuing the action until the specific reactions of the fibre disappear and a residue of cellulose is obtained.

(*f*) *Sulphite processes.*—For these the experiments must be conducted in sealed glass tubes or in a lead-lined digester. The fibre is sealed up with 7 or 10 times its weight of a solution of bisulphite of lime, and heated for 8 hours, raising the temperature gradually to $140°$ in the case of jute, or $160°$ in the case of woods. The products are thrown out into a dish and the insoluble cellulose filtered off, and the soluble products may be examined for the reactions described on p. 200. If neutral sulphite is used, take a 5 p.ct. solution of crystalline salt, and in this case raise to $160°$.

(5) **Furfural Estimations.**—The method used has been subject to extensive investigations, as stated in the text, the details finally adopted being those of Flint and Tollens (Landw. Vers.-Stat. 1893, 42, 381-407). The fibre is boiled with hydrochloric acid, of 1·1 sp.gr., in a flask attached to a condenser. The volume is kept constant by the addition of acid of this strength to the flask. The furfural which distils is converted into hydrazone in the neutralised solution, and estimated as such, with careful attention to all the precautions given in the paper above cited. The results are uniform, and the method is an important one for the student to master.

(6) **Methoxyl Determinations.**—The fibre-substance is boiled with concentrated hydriodic acid. Methyl iodide is formed, and is carried forward by a stream of carbonic acid through a special apparatus. The full details of the method are given in the original paper of Zeisel (Monatshefte f. Chem. 1885, vi. 989), and is also described in Vortmann's Anleitung z. Chem. Anal. Org. Stoffe. (Leipzig, 1891). This method is of growing importance in the investigation of vegetable substances, and should be thoroughly mastered by the student.

(7) **Nitration.**—The 'nitrating' acid is a mixture of equal volumes of nitric acid (1·5 sp.gr.) and sulphuric acid (1·83), previously mixed and cooled. The weighed quantity of the fibre-substance,

dried at 100°, is added to this mixture. The time and condition of treatment may be varied, and the influence of the variations noted. The product is removed from the acid and at once dropped into a large volume of water. It is then exhaustively washed until entirely free from soluble acid products, dried and weighed. The products may be analysed for nitrogen by the standard methods.

(8) **Ferric Ferricyanide Reaction.**—The solutions of ferric chloride and ferricyanide are prepared at normal strength, and mixed in equal volumes previously to the experiment. The fibre-substance is weighed and plunged into the red solution and allowed to remain, with occasional stirring. The deposition of the blue cyanide within the fibre-substance should be carefully observed under the microscope, and the gain in weight should be determined under varying conditions of digestion. The blue dyed product may be analysed in various ways to determine Fe and N.

(9) **Dyeing Operations.**—The fibre may be prepared by a preliminary boil in weak alkaline solutions. It should then be dyed up in the ordinary way, with coal-tar colours typical of the various groups. It will be found that the lignocellulose has a very varied dyeing capability, corresponding with the great variety of reactive groups which it contains. A systematic investigation of this question is very much wanted.

(10) **Ultimate Analysis.**—The elementary analysis of fibre-substance may be carried out by any of the standard combustion methods. In respect to the mineral constituents (ash), these must be determined in the usual way, by completely burning the fibre in a platinum dish, and an allowance made for the carbonic acid retained by the ash. With regard to the moisture, the fibre may be dried at 100–105°, and in transferring to the combustion apparatus, care must be taken that no absorption of moisture takes place.

The **Woods** may be treated in the same way as the lignified fibres, provided they are previously reduced to a state of the finest possible division. For this purpose the well-seasoned wood should be cut into shavings, with a fine plane. As stated in the text, there is much need for a thoroughly systematic examination of the woods comparatively with the typical lignocellulose, and there is also room for investigating those woods which contain colouring matters belonging to the aromatic group. In view of the more complex constitution of the woods, care must be taken, in bringing out the results, to distinguish between the fibre-substance proper and constituents easily removed by hydrolysis. Thus, to institute an exact comparison of the woods with the jute fibre, both should be taken after preliminary boiling out with dilute alkaline solutions under the same conditions, the residue being washed and dried after this treatment, and weighed in this condition for the several determinations. What is required is the comparative determination of the essential 'constants of lignification,' that is to say, elementary composition, hydrolysis numbers, cellulose, chlorine combining, furfural, methoxyl, nitration, and ferric ferricyanide reactions.

The Pectocelluloses.—This group involves the general methods of investigation of the carbohydrates of lower molecular weight, for the reason that the non-cellulose constituents are easily hydrolysed to soluble bodies by alkalis, and are then further broken down by acid hydrolysis to carbohydrates of definite and known constitution. In the examination of these compound celluloses, therefore, the methods of investigation to be found in standard works on the carbohydrates must be followed. The general scheme is that given in the text, p. 261 ; and as typical raw materials the following should be studied : flax, esparto, and fleshy parenchyma, such as found in the turnip, apple, pear, and similar fruits ; as bodies yielding the pectic acid series, and for the group of mucocellulose which yield the neutral carbohydrates (hexoses and pentoses), the investigations of Tollens (p. 223) should be repeated.

The Adipocelluloses.—This ground is in a very undeveloped condition, and much investigation will be required to establish the essential chemical features of this particular compound cellulose. It would be necessary to distinguish carefully between excreted by-products and the essential cuticular tissue. The most promising direction of investigation seems to be that resulting from previous treatment of the tissue by one of the sulphite processes. The drastic oxidations with nitric acid, and treatments with concentrated alkalis, are too severe to enable any conclusions to be drawn with certainty, from the products obtained, to the constitution of the parent molecule ; but a preliminary resolution into cellulose and non-cellulose, by a method involving a minimum of change, affords a much better basis for such investigations. As far as we know, this has only been investigated in a preliminary way, as stated in the text (p. 227), and an exhaustive study on these lines offers a most promising field of research.

INVESTIGATION OF RAW FIBROUS MATERIALS.

These various processes admitting of quantitative observation of results—the results being the constants of the individual fibres —having been studied with raw materials of known composition, and belonging to one or other of the groups, may be applied to fibres of unknown composition, and to complex mixtures, in which two or more of the compound celluloses are represented. The investigation of *fibres of unknown composition* proceeds on the following lines : (*a*) a general examination with reagents, to determine whether lignified or not ; (*b*) a general histological examination, determining its structural characteristics. A specimen is boiled for some time in caustic soda solution (1 p.ct. NaOH), then teased out, and the ultimate fibres measured. The length of the ultimate fibre is one of the most important criteria of value of a fibre. Cross-sections are cut and examined, to determine the general features of the fibre-bundle : the average number of fibres in the bundle, the dimensions, and the divisibility of the bundle. For a detailed account of these methods the student must read the special treatises on the subject (p. 243).

Proceeding with the chemical examination. If the microscope has revealed the presence of cuticular tissues, the raw material should be extracted with ether-alcohol in a continuous extraction apparatus, and the quantity of extract determined. The residue is treated for the estimation of cellulose by the methods previously described. The attendant reactions must be carefully observed. The percentage of cellulose is the most important of the chemical constants. The quality of the cellulose should be noted. It should be examined for resistance to further hydrolysis, and to oxidising agents, e.g. Fehling's solution, permanganate, &c. On these results, it is classified as of the cotton or normal group, the jute or wood cellulose group, or the esparto and straw group. The general character of the non-cellulose constituents will have appeared from their reactions ; further evidence is obtained by examining the solutions from the preliminary alkaline hydrolysis, and from the hydrolysis following the chlorination process. The latter should be carried out with certain precautions (see p. 96), and the chlorinated fibre may be washed, and the washings titrated, to determine the HCl formed.

The fibre may then be subjected to any of the special treatments, according to the group to which it is assigned.

Fibrous raw materials of more complex constitution must be examined with reference to external appearance, and the directions of application for which intended. If for the isolation of a textile fibre, the material is subjected to a prolonged boiling with sulphite of soda solution (2 p.ct. Na_2SO_3). The progress of the disintegration is watched, and when the adventitious tissues are sufficiently softened, the disintegration is aided by mechanical means—crushing, rubbing, &c.—and the cellular débris washed away in a stream of water. When perfectly cleaned, the proportion of fibre obtained is estimated. In the case of bast fibres, the entire bast may be stripped from the wood at an early stage, and the further purification proceeded with as described.

Where the raw material is to be examined for paper-making purposes—although it is possible to calculate with a fair amount of accuracy, from the quantitative data obtained on the small scale, the probable value of the material to the papermaker—it is advisable to carry out an experiment on a larger scale, under conditions similar to those which usually obtain in the papermaker's pulping process. For this purpose special apparatus is necessary. Thus, to make a complete examination, the material requires to be boiled under pressure with a quantity of caustic soda calculated from the data obtained on the small scale, and this must be carried out in a digester capable of resisting steam pressure up to (say) about 100 lb. per square inch. After the digestion the pulp must be removed, washed on a wire gauze filter, and then put through a quantitative bleach operation. For this purpose the pulp should be well broken up, placed in about 30 times its weight of water (estimated), and bleaching powder solution is added, calculated to (say) 20 p.ct. on the weight of the pulp obtained, which will be approximately known from the cellulose estimations by the laboratory methods. When the material is bleached, an aliquot portion of the residual liquor is drawn off, and in it the residual hypochlorite is estimated by the usual methods. The difference gives the amount actually consumed in bleaching the pulp. The pulp is then washed off and treated with a little antichlor (sulphite of soda), and again washed. It is then pressed up in suitable moulds, which may be easily made by attaching perforated zinc to square frames of wood. Further than this, if it is required to make an actual paper-making experiment with pulp, this must be prepared by beating in a model beater, and then converted into sheets by the hand process. The details of such manipulation are, of course, highly technical, and for organising such a plant and process, the assistance of an expert would be required. Having determined the yield of cellulose by the laboratory method, and ascertained its characteristics, the yield by the alkali boiling and bleaching process, and the proportions of caustic soda and of bleaching powder required for the isolation of the pure cellulose, complete data are at hand for valuing the raw material by comparison with staple materials of the same class.

The application of these methods to the investigation of green plants in physiological investigations, or to fodder plants, green or otherwise, is a province in which it is difficult to lay down definite schemes. The choice of method must depend largely upon the subject to be investigated, and for the present—that is to say, until our knowledge is more complete—the selection must remain more or less arbitrary. The following general considerations will serve as guides in selecting methods suitable for particular inquiries. Thus in green plants it is important to distinguish between what we may call 'permanent' or fundamental tissue, and cellulose or lignocellulose. The fundamental tissue might be defined as the assemblage of cells which constitute the plant, or part of the plant, *less* the cell contents, including all excreted products. Therefore, to isolate such a complex we must proceed by way of selecting reagents calculated to remove particular constituents, or groups of constituents, with the least action upon the cell wall, or cell substance proper, of whatever kind. In investigation of the permanent tissue of the Gramineæ which the authors are prosecuting, the following process is used for its isolation :

(1) The material is exhausted with boiling alcohol.

(2) It is digested for 6 hours in cold dilute caustic soda (1 p.ct. NaOH). It is washed off from this solution, first cold and then boiling hot.

(3) It is digested for some hours in cold dilute hydrochloric acid

(1 p.ct. HCl), and again washed off cold and hot. The residue from this treatment is defined as 'permanent tissue.'

If, now, a plant or plant-substance were to be investigated containing a large portion of starch, it would be necessary to precede these hydrolytic treatments by a process acting selectively on the starch, viz. the substances reduced to a fine state of division, boiled for a short time with water, left to cool, and treated with malt extract, being digested for some hours at the most favourable temperature for conversion. After this, which should follow the alcoholic exhaustion, the remainder of the processes may be proceeded with in order. (Compare V. Stein, Exper. Stat. Record, 5, 613, from Ugeskr. f. Landmand, 39, 706.) Such a residue will contain a certain proportion of ash constituents and nitrogen, for which, in certain cases, allowance must be made, by the usual methods of determining and calculating. The difference between this product and that known as 'crude fibre' (Weende method) will be noted. The important aspect of these methods and their differences is appreciated in dealing with a complex such as the constituents which yield furfural. Of these, the product known as wood gum (pentosan) is soluble in dilute alkaline solutions in the cold, and would be eliminated under these treatments; but the furfural-yielding constants are only partially eliminated by the treatment, even from non-lignified tissues. But, in the mean time, it is not safe to follow on the older lines, which usually were held to sharply divide group from group. We may affirm generally that no hydrolytic process can effect any such separation, and this is particularly to be noted in regard to the furfural-yielding constituents. It is advisable, therefore, to bear in mind that any process selected is more or less arbitrary, and gives results which, while they may be perfectly valid under conditions of strict comparison, are not to be interpreted outside these comparisons except with reservation. This will be specially appreciated when the results of the proximate analyses are taken as evidence of feeding value. This entire subject is very much in need of revision, and we hope that both the theoretical matter and the experimental methods described in this treatise will contain suggestions of methods by which these problems can be more effectively solved.

The Analysis of Textiles and Paper.—The various processes of quantitative determination that have been described are available for the examination of fibrous mixtures, with a view of determining their composition. In textile fabrics this matter seldom arises, unless in the broader distinction of the vegetable from the animal fibres. The subject is exhaustively treated by H. Schlichter (J. Soc. Chem. Ind. 1890, 9, 241). The vegetable fibres are separated as a group by the process of boiling with the alkaline hydrates (5-10 p.ct. solution), which dissolves the nitrogenous animal fibres, leaving the vegetable fibres not, of course, unacted upon, but sufficiently so to enable the method to be regarded as a quantitative separation. The distinguishing of the various vegetable fibres, thus separated, from one another is only possible by microscopic observation with the additional aid of reactions. No general directions can be given for investigations of this character. They involve experience of histological methods, and acquaintance with the characteristics of minute structure, and of the special features which render their quantitative estimation possible within a sufficient approximation for all practical purposes. The examination of papers involves the identification of the vegetable fibres only. The composition of a paper is first generally indicated by its appearance. It is only in the mixed class of white papers that investigation has to be carried out in minute detail. The following are briefly the methods adopted, so far as these are chemical :

(1) The paper is treated with aniline sulphate solution in the cold. The presence of mechanical wood pulp is indicated by the yellow stain produced, and the proportion approximately by the depth of colour. Many of the ground woods, it may be remarked, yield a pulp of sufficient whiteness to be used in what may be called white paper, and is frequently present in 'white' and 'toned' printing papers.

(2) The paper is boiled with the solution of aniline sulphate. The presence of esparto and straw 'celluloses' is indicated by the characteristic rose-red reaction. Papers giving no colour reaction with aniline sulphate are probably composed of rag fibres (cotton linen), with or without bleached wood cellulose. It is evident that the approximate composition of a paper is thus very quickly determined by its chemical reactions ; but it is often necessary (3) to make quantitative estimations within narrower limits, and for these a microscopic investigation is at present the only method to be recommended. The examination is, of course, facilitated by taking

advantage of chemical reactions to differentiate the fibres, the method then consisting in the approximate estimation by actually counting the fibres visible in the microscopic field according to their identity, and averaging the results over a sufficient number of separate examinations, and, where possible, by separate observers. This, again, is a species of investigation which requires considerable experience, which cannot be communicated in the form of working directions.

(4) For actual quantitative work, of course, any of the reactions of which full details have been given in the earlier sections of the work are available. Thus, for instance, in white paper, found as above to be composed of rag fibres and celluloses of the Gramineæ only (esparto and straw), it will be evident that, as the latter yield 12–14 p.ct. of furfural on boiling with hydrochloric acid, and the former at the outside 0·5 p.ct., a furfural estimation in the usual way (p. 99) would give a close approximation to the proportions of the two groups of cellulose. In the case of mechanical wood pulp, if the proportion is high, there are two or three reactions available as a quantitative estimation. First, the statistics of chlorination according to the methods described on p. 104. Secondly, estimation of furfural; but this is only available in the absence of celluloses of the Gramineæ. Thirdly, methyl estimations; which, again, depends on the ascertained absence of other fibres also containing this group. Fourthly, the colour reactions with derivatives of p-phenylene-diamine, as described on p. 174. And, lastly, the elementary analysis might even be made and taken as a basis for calculating the proportion. These brief notes will be sufficient to show the student how to set to work in the laboratory to examine these particular mixtures of fibres with the help of the reactions previously described in detail.

Principles of Cellulose Technology.

Following these notes of laboratory and general experimental methods, we shall briefly discuss the applications of theoretical principles and deductions to the practical processes of the arts. The celluloses and compound celluloses are familiar to us in multitudinous forms, both 'useful' and 'ornamental'; and the processes by which they are manufactured, or treated for various purposes after being manufactured, involve the special chemistry of the raw materials at every turn. It must be confessed that the arts of spinning, weaving, bleaching, and dyeing have been highly developed upon a very slender chemical foundation so far as regards the raw materials themselves. There is no doubt, on the other hand, that an ample field for technological developments will be opened up by the systematic application of the more definite chemical knowledge now available. It may in fact be affirmed as a general principle, established also by long and invariable experience, that there are no results of chemical investigation, however recondite they may appear, which are not in their due order absorbed into the province of technology. As it is the province of the technologist to give a complete account of his processes in terms of the factors which contribute to the result, it will be very evident, from the ensuing discussion of cellulose technology, that much remains to be done before the industries of fibre-preparing, spinning and paper-making, bleaching, printing and dyeing, can be said to rest on such a basis.

Preparation of Fibres from Fibrous Raw Materials. Processes with this object divide themselves into two groups: (*a*) for the separation of *spinning* fibres, (*b*) of paper-making fibres. While the latter are almost exclusively chemical, the former are as exclusively mechanical, and require therefore but a brief general notice. (*a*) The spinning fibres are mostly obtained from annual growths. With the exception of cotton, which is a seed-hair, they form part of complex structures, and are themselves either localised into a special tissue (bast fibres, see Appendix, figs. 1–4), or scattered more or less irregularly (fibro-vascular bundles of monocotyledons, see figs. 5, 6). Structurally they are differentiated, as fibres or elongated cells, from the cellular tissue by which they are surrounded, the component cells of which are spherical or cubical, with more or less elongated deviations. Chemically the 'fibres' are differentiated by their superior resistance to the attack of hydrolytic agencies. It is a matter of common observation that 'fleshy' structures are more perishable than fibrous. The chemical constitution of the tissue-substance of this less resistant order has been only superficially investigated; generally the parenchyma of flowering stems may be classed with the pectocelluloses. Where the fibrous raw materials are subjected to a preliminary treatment, with the object of facilitating the separation of the fibres from non-fibrous tissue, it is always a process of hydrolysis, and usually the 'natural' or spontaneous process of fermentation.

Thus *flax* and *jute*, to select the prominent types, are treated by the process of retting or steeping. This consists in submerging the stems in stagnant water; a spontaneous fermentation is set up, with the result that the less resistant (cellular) celluloses are disintegrated and broken down. In the case of flax the retted 'straw' is dried off, still containing the fibre. This is separated by the mechanical process of breaking and scutching. In the case of jute the bast layer is separated at once from the retted stem, by the manual operation of stripping; and freed from cortex and adhering residues of parenchyma, by beating the strips upon water. It may be stated generally that these processes have not been systematically studied with the view of localising the effects produced. An investigation of the subject with the more precise methods of diagnosis now available would be a most valuable contribution to theoretical and industrial science.

As an illustration of the desirability of more precise information, the history of the attempts to substitute the natural by artificial processes, in the case of flax, may be cited. Various chemical treatments of the stem or straw have been proposed, and indeed worked. The yield of fibre in this plant being relatively high (18-23 p.ct.), and the value of the fibre being also high (40*l.* to 60*l.* per ton), such treatments are not precluded on economic grounds. Moreover, as the natural conditions most favourable for retting are not to be counted on in the capricious climates of temperate regions where the flax is chiefly grown, an artificial process admitting of exact control is very much to be desired.

The difficulties to be overcome are not so much those of separating the fibre as of separating it in a condition as favourable for spinning as the product of the natural process or processes.

The analysis of the fibre shows that, in addition to the pectocellulose or fibre proper, there is present an unusual proportion of oil-wax constituents (3-4 p.ct.). It appears from later investigations of the spinning process (*infra*) that the 'natural' balance of these constituents constitutes the 'optimum' of spinning properties. All the artificial processes, which are usually treatments with hot alkaline solutions, disturb this balance, removing both pectic and oily constituents. Moreover, the oils found in the 'natural' product are in part *produced* in and by the retting process; and the pectic constituents of the fibre are present, not only in different proportion, but in different condition chemically. As a matter of history, these processes have failed technically—i.e. in producing a fibre with the high spinning qualities of the ordinary product— and with commercial results more or less disastrous. Had investigators based their labours upon the natural model, as defined by exact chemical investigation, such failures would have been obviated.

But investigation is still needed to elucidate (1) the changes produced in the oil-wax components during retting; (2) the effect of the retting process upon the pectic constituents of the fibre proper. Upon the results of such investigation it might be possible to devise an artificial process giving similar results, but the balance of conditions to be observed is necessarily one of very fine adjustment. Any artificial treatment hitherto attempted resembles the natural in being a process of hydrolysis ; the reagents to be used have been of the alkaline group, and employed at relatively high temperature -- conditions which make it extremely difficult to limit and regulate their action.

The authors have made investigations of the 'retting' action of dilute solutions of sodium carbonate, silicate, and sulphite comparatively with the soda soaps, and with the natural process. Of these several reagents the action of the soda soaps alone resembles that of the natural process, the 'straw' thus treated behaving in the scutching process very similarly to the ordinarily retted product. But the scutched fibre is of inferior spinning quality owing to the partial removal of the pectic and oily constituents.

Treatments of spinning fibres, after removal from the plant, are sometimes resorted to in order to improve the working qualities of the fibre in the mechanical processes of refining and drawing preparatory to the actual spinning process. The great desiderata in a yarn are uniformity and strength, and yarns are valuable in proportion to fineness. The spinning unit in all the vegetable fibres, with the exception of cotton (and perhaps rhea), is a complex or bundle of the ultimate fibres. In the processes of hackling and drawing, it is sought to reduce or divide the bundles to the maximum of fineness. In flax the subdivision of the bundles is carried very far and without auxiliary treatment. In jute, on the other hand, the bundles are much more firmly compacted ; and, as a lignocellulose, it possesses none of the 'gummy' properties of the pectocelluloses, and is also relatively deficient in oily constituents. This fibre is subjected therefore to a preliminary treatment with oily aqueous mixtures of varying composition, the incorporation of which greatly improves the spinning qualities. The treatment and its effects are, however, rather mechanical than chemical. Hemp has been chemically treated for the same general purpose, by a process devised by the authors, consisting in a digestion of the fibre with dilute solutions of basic sodium sulphite at high temperatures. This enables the fibre to be drawn and spun to finer numbers, and has proved especially valuable in the manufacture of the finer counts of shoe-threads.

Rhea, or China grass, is a fibrous material that also requires chemical treatment preparatory to spinning, but for a different reason. This fibre is separated from the mature stems by stripping the entire bast and cortex from the wood. The ribbons thus obtained are treated by various processes for the removal of what is commonly termed the 'gum.' The pectic constituents of the fibre and parenchyma readily yield to the action of alkaline solutions, and the disintegrated cellular residues are then easily removed by mechanical operations.

The process of purification is, in the case of this fibre, carried to the full extent of isolating a pure cellulose. The ultimate fibre, being of the unusual length of 40-200 mm., is a spinning unit of sufficient dimensions, comparable with the flax *filament*. Its spinning qualities are, on the other hand, inferior to those of flax, and it is probable that much better results would be obtained with this fibre by spinning it in a condition more nearly that in which it occurs in the plant.

Generally it may be said that the chemical treatment of these fibres preparatory to the mechanical operations of the spinner has been investigated on purely empirical grounds. There are a number of questions of both theoretical and practical import which await systematic inquiry. The purpose of this superficial and general discussion of the subject is to indicate some of the directions in which the theoretical conclusions arrived at in the earlier sections of this work may be applied.

SPINNING PROCESSES.—The various spinning processes for converting into yarn the fibres obtained as above described are for the most part purely mechanical operations. They depend in an important way upon the minute structure of the spinning unit, whether that is an ultimate fibre (cotton, rhea cellulose) or a complex filament (flax, hemp, jute, &c.) ; and therefore indirectly upon the chemical properties of the fibre-substance. These questions are exhaustively treated by Vétillart in his work upon the Vegetable Textile Fibres. There is one process only which directly involves the question of the chemical composition of the fibre-substance, and that is, the 'wet process' of flax spinning The history of flax spinning shows three periods of development : (1) At first the fibre was spun dry in the same manner as jute is at this day. (2) It was found that the drawing properties of the fibre were much improved by maceration in cold water, and the wet spinning enabled the fibre to be spun to much finer qualities of yarn. (3) A still further advance was made by the introduction of hot water, this treatment taking place on the spinning frame, the roving running through a trough of water kept at 50-60°, and receiving its final drawing and twisting immediately as it emerges from the trough. This may be considered the universal process of spinning fine flax line yarns at the present day.

There have been numerous attempts to realise a still further improvement by alkaline treatments of the most varied kind, either on the spinning frame itself—i.e. by adding the alkaline reagent to the 'spinning trough'—or by a previous treatment of the roving. Such processes, however, have only come into limited use. A more successful attempt to still further raise the spinning qualities of flax is that of C. C. Connor, of Belfast, who patented in 1888 a process based upon the results of the authors' investigations of the constituents of the fibre. From these results it appeared that a considerable proportion of the oily constituents were of a ketonic character, and were readily emulsified by treatment with solutions of such salts as sulphite and phosphate of soda.

The addition of such salts to the 'spinning trough' might be expected to bring about a much more perfect distribution of the oil-wax components throughout the fibre-substance than is possible by treatment with hot water. The pectic constituents, also being further attacked than by hot water, might be expected to be brought into a more favourable condition for yielding to the drawing action of the frame. Experience has verified these predictions, and the working of the process on the large scale has shown conclusively that in the coarser Russian flaxes there existed an undeveloped margin of spinning quality which is fully realised under the new process. As it has also been shown that the weight of yarn spun from a given weight of roving is not sensibly different from that obtained by the ordinary hot-water process, it is evident that these alkaline salts, under the conditions adopted, do not exert any undue solvent action on the constituents of the fibre, but are limited in their action to bringing about the optimum condition for drawing and subdividing the fibre-bundles. The salts are used in the process in the form of 1-2 p.ct. solution, the proportion being adjusted to the quality of the flax.

(*b*) The paper-making fibres are obtained from very various sources, largely from the rejections of the spinning and weaving industries (scutching tow and waste, jute butts, spinning wastes, rags and cuttings of all kinds). In addition to these there are a number of vegetable raw materials which are treated directly for conversion into fibre or pulp—e.g. esparto, straw, wood.

The processes of treating wood have already been discussed, as they admit of classification on strictly theoretical lines, and afford a useful illustration of the general principles of the relation of the cellulose to the non-cellulose constituents of the compound celluloses. In extending this classification to the wider range of raw materials above indicated, it is necessary to remember the general features of the three groups of compound celluloses, and more particularly the conditions under which they are resolved into cellulose and non-cellulose, observing also that any process of resolution to be available for the purpose in question must be limited in its attack as much as possible to the non-cellulose components. The following may be laid down as a broad principle of economy in such treatments: effects required to be produced should be separately accomplished, and obtained by specific reagents. It must be conceded at once that this is an ideal seldom realisable. The treatments of the papermaker are nearly always 'overhead' treatments, in which one process and one reagent is employed to work a very complex mixture of chemical decompositions. But because practice tends to stereotype itself on the lines of apparent simplicity, it is not for the chemist to accept this order of things as unassailable. Experience has shown, and is continually showing, that 'division of labour' in reactions is as economical as it is in other branches of work; and it is a particular purpose of this discussion to suggest a careful revision of these 'overhead' treatments, with the view of improving methods wherever possible.

The pectocelluloses from our present point of view need no discussion. They are easily resolved by alkaline hydrolysis of the simplest kind, i.e. boiling at the ordinary boiling temperature with solutions of the alkalis. The resulting cellulose would be approximately pure and structurally disintegrated, i.e. in the condition of ultimate fibres.

The lignocelluloses present problems of a totally different character. The more resistant members of the group—viz. the woods—are 'pulped' by various processes which have been already described. Some of these depend upon a specific attack of the non-cellulose constituents, whether by way of synthesis with the reagents employed (sulphite processes) or radical decomposition (nitric acid process); others may be rather described as 'overhead' treatments, in which a highly complex series of chemical changes are determined which are by no means confined to the non-cellulose constituents, but affect the cellulose also, and prejudicially in regard to yield. Such are the alkali processes. The typical lignocellulose jute stands on a different footing from the woods. The latter are used either (1) as 'mechanical wood pulp,' obtained by merely grinding the wood; (2) as 'chemical pulp,' which is a more or less pure wood-cellulose, obtained by the processes previously described. Jute, on the other hand, is largely, in fact chiefly, used as a disintegrated and purified *lignocellulose*, 'pulped' by a process which leaves the cellulose and non-cellulose still in intimate combination. The process giving this intermediate product is that of boiling with lime at relatively low temperatures ($105-115°$). It is applied to the rejected root ends ('cuttings'), which contain also 'pectic' (incrusting) constituents and residues of cortical and past parenchyma. These are resolved by the treatment, and the fibre-bundles of the lignocellulose proper are largely disintegrated, a certain proportion being also hydrolysed and dissolved. The product (pulp) is therefore a purified lignocellulose, in a condition easily yielding to the subsequent mechanical operation of beating.

Jute may be treated for the isolation of a *jute cellulose* by any of the processes described for the woods. A process also used to some extent is that of *chlorination*, the fibre being first prepared by boiling in a weak alkali, and, after washing, exposing in closed chambers to an atmosphere of chlorine gas, afterwards removing the chlorinated product by again boiling in alkali. This process is the laboratory method of isolating cellulose applied on the large scale. In the form of cellulose, however, jute comes into unfavourable competition with the woods, and the process is therefore not much used.

The *adipocelluloses* come into consideration in this connection, merely as adventitious tissue-constituents. They are a source of considerable difficulty on account of their resistance to the attack of reagents. They occur, of course, chiefly in such raw materials as are entire stems or leaves, e.g esparto and straw, and are characterised by admixture with chlorophyll (esparto) and oil-wax constituents. For the removal of the latter the alkaline treatment is relied upon, the conditions of which for the purpose require to be much more severe than for the non-cellulose of the fibre-constituents proper. The cuticular cells themselves are only slightly attacked by the process, and are obtained in the pulp, in which they are easily recognised under the microscope by their very characteristic form. The neutral waxes are obtained at the end of the boiling process in mechanical mixture with the mass of pulp and liquor, and they collect on the surface of the lixiviating vats used in the continuous process of washing esparto pulp. The treatment of such raw materials, in which *all* the compound celluloses are represented, is perhaps the best illustration of what has been expressed by an 'overhead' treatment. The ordinary processes are, in fact, a crude aggregate effect, and it is more than probable that means may yet be devised for more specific treatments, in harmony with the broad principle of economic chemical work. In all these industrial processes for isolating cellulose, moreover, the yield is considerably less than what may be considered the theoretical, i.e. the proportion isolated by the method of chlorination (p. 95). For these raw materials which have been more especially mentioned, the following may be taken as the comparative yields :

Laboratory method. (Yield of dry cellulose on dry raw material.)		Papermakers methods. (Yield of air-dry pulp on air-dry raw material.)
Esparto	50–55 p.ct.	43–47 p.ct. Alkali process
Straw	50–55 ,,	33–37 ,, ,, ,,
Wood	50–55 ,,	35–43 ,, ,, ,,
		42–48 ,, Bisulphite ,,

It is obvious, therefore, that the cellulosic constituents of the fibres are considerably attacked, and that there is an ample margin for improved results in regard to quantity as well as quality of the fibre produced (pulp).

BLEACHING PROCESSES.—These processes appear to divide themselves into the two groups : (a) the bleaching of textiles ; (b) of paper pulp. It will be evident, however, from the present treatment of the subject, that bleaching is a process of purifying a cellulose or compound cellulose from adventitious constituents, whether mechanically mixed with the tissue or fabric, or chemically united to the ultimate fibre-cellulose ; and on this view of the subject bleaching treatments divide themselves into (i) processes for the *purification of a compound cellulose,* with removal of colouring (or discolouring) matters (jute textiles ; flax yarn, 'creaming' and half bleaching process ; linen textiles, part bleaching ; pulps for wrapping and coloured papers) ; (ii) processes for the *isolation of a pure cellulose* (cotton textiles, linen textiles, papermaker's cellulose).

The *bleaching process proper* is the whitening or decolourising process which follows such alkaline treatment as those already described. The bleaching is invariably a treatment with oxidising agents, usually alkaline ; 'bleaching powder' or calcium hypochlorite is the 'staple' reagent. Other hypochlorites (sodium and magnesium), obtained by double decomposition from the former, are largely used, and oxidising solutions obtained by the electrolysis of solutions of the chlorides (chiefly $MgCl_2$) are also now extensively used (Hermite process).

Chemically, therefore, the bleaching processes of the arts consist essentially of the two treatments: (1) alkaline hydrolysis followed by (2) alkaline oxidations.

In the processes of the *first* group the alkaline treatments are of the milder order, the purpose being to dissolve and remove the minimum of non-cellulose constituents, consistently with obtaining a uniform and sufficiently high colour (bleach) in the finished product. As therefore a large proportion of the more oxidisable (non-cellulose) constituents is retained in the pulp or fabric, the consumption of the bleaching agent in the after process is relatively high. It is in fact used up, not in selectively oxidising those constituents which are the colouring matters of the alkali-boiled fibre or fabric, but obviously in a general oxidation of the non-cellulose constituents in the order of oxidability.

Two processes may be considered as typical of this group:

(1) *Jute fabrics and jute pulp.*—Jute itself may be whitened considerably by regulated oxidations. In the case of this fibre, however, it is difficult to control the action of bleaching powder. The avidity of the lignocellulose for chlorine is such that should any free hypochlorous acid be formed in the solution, chlorination of the fibre immediately results. The presence of the lignone chlorides in the fibre is a source of considerable danger. Being unstable they are gradually decomposed, with liberation of hydrochloric acid, which rapidly disintegrates the fabric. The neglect of this property of the lignocellulose has led to disastrous consequences in manufacture. An industry established some years ago for the bleaching and printing of jute cloth was ruined through the wholesale 'tendering' of the goods from this cause. The process adopted consisted in (*a*) boiling in weak alkaline solutions

(carbonate and silicate of soda), (b) bleaching with calcium hypochlorite solution in a closed vessel (Mason Kier); after which the cloth was washed, 'soured' in weak acid, washed up, and dried. The printing processes were those ordinarily employed for cotton goods, the colours being developed and fixed by the usual process of steaming, in an atmosphere of dry stream at 4 lb. (per square inch) pressure. It was in the latter case that the discolouration and tendering effect chiefly showed themselves. The cause being traced, the remedy was easily devised, the process being modified as follows: (a) in the bleaching process, sodium hypochlorite was substituted for the lime compound—and in this way the chlorination of the fibre-substance was arrested; (b) as a last treatment, after souring and washing, the goods were run through a solution of sodium bisulphite (1 p.ct. SO_2), and dried after squeezing. In this way a residue of the normal sulphite (Na_2SO_3) was left in the cloth, and this was found to prevent discolouration in the steaming process.

In this method of bleaching, the loss of weight of the fabric was from 8–12 p.ct., the colour obtained being the pale cream shade of the highly purified lignocellulose. The results obtained by bleaching with permanganates are superior to those with the hypoch'orites, but at much greater cost. The process is therefore but little used industrially.

(2) *Linen yarn and cloth: partial bleach.*—In the linen industry, in addition to the full bleaching of shirtings, sheetings, cambrics, &c., there is a large practice in partial bleaching of various grades. These processes are familiarly designated as 'whitewashing,' in contradistinction to the 'bottom bleaching': in the former the non-cellulose constituents are only partially removed, and the residues whitened by bleaching agents; in the latter they are entirely eliminated, leaving the residue of pure flax cellulose. The partial bleaches in question are

obtained by a light alkaline boil, followed by a treatment with bleaching liquor (hypochlorite), these treatments being once or twice repeated for higher grades of bleaching. The consumption of bleaching powder is relatively large (10–30 p.ct. of the weight of the goods), a considerable proportion being used up in oxidations which do not contribute to the bleaching effect proper. The processes are therefore not economical in the strict sense of the term, and are capable of considerable improvement in the direction of a more specific attack of the coloured constituents of the yarn. In various grades of paper making, also, similar half-bleaches are practised.

Jute (cuttings and waste) is boiled in lime and bleached with bleaching powder solution, the resulting pulp being of a yellow to a yellowish-white colour, still retaining a large proportion of the non-cellulose constituents of the original fibre, and giving all its characteristic reactions. Flax wastes (scutching tow) are boiled with lime or soda to soften and disintegrate the residues of wood (sprit), and the pulp is bleached with hypochlorites.

The principle of these treatments is, however, one and the same for all, and is sufficiently illustrated by the examples discussed.

(*b*) The second group of bleaching processes, of which the goal is a pure cellulose (or oxycellulose), differ from the above in this general and important particular : the chemical work is thrown chiefly on the alkaline boiling processes, the bleaching treatment proper being limited to the oxidation of the coloured residues from these treatments. Thus in cotton bleaching, while the consumption of caustic soda may be taken at 80–100 lb. per ton of cotton goods, the bleaching powder required is less than 30 lb. per ton, a proportion of which is wasted in the unavoidable losses attending the washing away of residual liquors. In both cotton and linen bleaching of this order, moreover, the bleaching solutions are used in a highly dilute form (0·5-2·0 p.ct. bleaching powder). In papermakers' cellulose bleaches, while it is true that by far the greater proportion of the chemical work of purification is thrown upon the pulping process, the consumption of bleaching powder in the bleaching process proper is in some cases considerable. In the bleaching of rag pulp (cotton and linen) the average consumption is from 2-5 p.ct.; in straw and esparto pulp, 10-15 p.ct.; and sulphite wood pulp, 15-25 p.ct. In these latter cases we have a further illustration of 'overhead' treatments—i.e. in order to produce a certain result in a given time and a single process, a large amount of waste energy is expended. These celluloses are, as we have already seen, very different constitutionally from the normal type: they are easily hydrolysed, and in the alkaline bleach liquor a considerable further proportion of the fibre-constituents are dissolved and undergo oxidation of a perfectly useless character. To minimise these wastes of the oxidising agent, the practice of intermediate washing is sometimes resorted to; and by thus separating the effects of hydrolysis and oxidation, the latter is controlled into the directions of useful, i.e. bleaching oxidations. The economy of bleaching powder which results is very considerable, and it is not a little remarkable that so rational a plan is not more generally adopted.[1]

Of the textile bleaches of this group there are two which may be selected to illustrate general principles, viz. the cotton bleach and the linen full bleach.

In COTTON-CLOTH BLEACHING the most important process is the alkali boil. The treatment is varied to suit the great variety of goods which undergo the process, but for our present

---

[1] A very thorough treatment of papermakers' bleaching processes will be found in Griffin and Little's 'Chemistry of Paper Making' (1894), chap. v. pp. 275-300.

purpose we need consider but the one in which caustic soda is used. With this reagent, in the form of a 1-2 p.ct. solution of NaOH, cotton goods are effectively cleared of their non-cellulose impurities in a single treatment. The conditions of the process are : (1) a saturation of the goods with the alkaline lye, usually effected by passing the goods in continuous length through the hot liquor, removing the excess by squeezing, and piling up in the 'kier,' or boiling-vessel ; (2) the boiling process, in which the goods are subjected to the further action of the alkaline lye at temperatures of 105-115°, and under corresponding steam pressures. The liquor is kept in circulation through the goods, and the 'boiling' is continued from six to ten hours.

After this treatment the goods are washed free from the alkaline lye and the dark coloured soluble products of the action, and are then of a greyish-brown colour. The residual impurities are then removed in the bleaching process proper, which consists in exposing the goods to the action of bleaching powder solution. The goods are then washed and 'soured,' to remove basic residues. This round of operations is sometimes repeated, though with weaker solutions, in the case of heavy goods, or of goods made of the more refractory Egyptian cottons, which contain a red brown colouring matter. The process, however, need not be followed into its technical details. It is one of great simplicity, and aptly illustrates the resistance of the normal cellulose to alkaline hydrolysis and oxidation under somewhat severe conditions. The fibre itself loses from 7-10 p.ct. in weight under the treatment. The products removed in solution have been investigated by Dr. E. Schunck, who resolved the dissolved products into (*a*) *Cotton wax*, a neutral wax, melting at 80-86°, having the composition C 80·3, H 14·4 ; (*b*) *Fat acid*, which appeared to be a mixture of palmitic and stearic acids. The analytical numbers obtained were C 75·5, H 13·0. (*c*) *Pectic acid*, a gelatinous acid body, having the composition and properties of the acid described by Frémy. (*d*) Two colouring matters— (1) soluble in alcohol, (2) insoluble—having the following composition :

	(1)	(2)
C	58·48	57·7
H	5·80	6·05
N	5·30	8·74

(Mem. Lit. and Phil. Soc. Manchester, [3] 4.)

In addition to these substances, which are constituents of the fibre proper—including residues of cell-contents—the alkaline treatment breaks down the residues of the seed envelopes (motes) which survive the mechanical operations of preparing, and find their way into the yarn. The proportion of these by weight is, however, relatively insignificant, though they are a source of some difficulty to the bleacher.

There can be little doubt that the cotton cellulose undergoes certain molecular changes during the process of a normal bleach. From what we know of its constitution and reactions we may affirm that it does not remain inert under treatments of this severity ; but our methods are not sufficiently refined for differentiating the product from the cellulose as contained in the raw cotton. There is perhaps one exception to be noted, which is, that the bleached cotton yields from 0·2-0·6 p.ct. of furfural on boiling with hydrochloric acid, which may be taken as an indication of the presence of a small proportion of oxycellulose.

As already pointed out, cotton is very easily oxidised to oxycellulose under the joint action of calcium hypochlorite (in dilute solution) and carbonic acid. The researches of Witz, who established the general conditions of these oxidations, were carried out at a date (1882 85) when there were none but qualitative reactions (dyeing phenomena &c.) available for demonstrating the formation of oxidation products. As it is probable that condensation to furfural is a property of these oxycelluloses, and the estimation of this product is reduced to a method of precision, it would be important to investigate the cotton in three stages, viz. : (1) in the raw state ; (2) after alkaline treatments of varying degrees ; and (3) after bleaching processes of various kinds and degrees, for the presence of furfural-yielding constituents and their quantity.

The classification of cotton-cloth bleaches into 'market bleach,' 'madder bleach,' &c., involves no important question of principle ; and for description in detail of the variations of treatment practised in the several grades, the technological textbooks must be consulted. We would specially mention, in passing, the article on 'Bleaching,' in Watts' Dictionary (Applied Chemistry, new edition), which gives an excellent survey of the history of development of the art. It may very well be assumed by those familar with this history that we have arrived at terminal excellence in the art. From the economical point of view it is, perhaps, difficult to see any unexplored margin. But, on the other hand, there is evidence of important recent progress in a direction of improvement, which will be evident from the following considerations. A web or fabric of cotton must be always considered by the technologist from the point of view of minute structure, the structure being that of the ultimate fibre, complicated by the spinning twist and the interlocking of the yarns in the weaving. The penetration of cotton goods in the mass by liquid reagents is obviously a highly complicated process. In the first place, complete penetration is probably possible only by previous exhaustion of the air contained in the tubes ; and, secondly, penetration of the substance of the cell wall must involve osmotic phenomena. Osmosis is complicated in two directions : first, by the filtering-out of the active reagent employed in the treatment; and, secondly, by physical changes in the cotton itself or its non-cellulose constituents. In the alkaline treatments of cotton it is of importance that the action of the alkali, water, and heat should be as nearly as possible equal and simultaneous throughout the mass. The advance of the caustic alkali process over the successive treatments with lime and soda ash of the older methods consists chiefly in this, that by the more rapid action of the more powerful alkali, secondary changes of the more oxidisable non-cellulose constituents are reduced to a minimum; and these are dissolved away by a single operation, with a minimum residue of products to be removed in the bleaching process proper. In the ordinary processes of bleaching, the result attained is simply measured by the appearance of the cloth. The printer, however, requires something more than a good white. The operations of calico printing in many cases involve a dyeing process, not of the whole cloth, but of the design or pattern printed with suitable mordants, the cloth itself being required to resist the colouring matter of the dye bath. Many 'market bleaches' are therefore very inferior in point of purity of the cellulose to the 'madder bleach' of the printer, and will dye up with alizarin and similar colouring matters, which the latter will resist under the same conditions. It is in regard to this important distinction, and the further refinement of the bleaching process for the 'madder bleach,' that progress continues to be made.

Linen bleaching.—The full bleach of flax goods, which consists in the isolation of the pure cellulose, is a much more complicated process than the bleaching of cotton-cloth, though based upon identical principles, and involving for the most part precisely similar methods.

The proportion of non-cellulose constituents in flax is very high, varying from 20-35 p.ct. of the weight of the fibre, according to the conditions of growth and the methods of separating and preparing the fibre. The greater proportion, being pectose-like substances, are easily attacked by alkaline hydrolysis; but the removal of a large weight of such products from a mass of cloth is not an easy operation. The alkaline treatments are therefore graduated, and are three or even four times repeated before the cloth is considered ready for the bleaching treatment. There are then the additional complications of the wood residue (sprit) and cuticular constituents which very much protract the after processes, or bleaching proper. These processes may be divided into series, the first of each series being the process of treating with dilute solutions of the hypochlorites. These involve prolonged exposures (6–12 hours), the cloth being entirely submerged in the solution. After this follows usually the souring process, and to this succeeds a light boil in progressively weaker alkaline solutions. These treatments, with intermediate washings, constitute the 'round.' After each round, the cloth, or rather the residue of non-cellulose constituents, is in the most favourable condition for the further attack of the oxidising or bleaching agent. These processes are repeated until the impurities are finally eliminated. In addition to these treatments, which are those practised by the cotton bleacher, linen undergoes the process of 'grassing,' i.e. is spread out upon grassfields and exposed for one or two days to the action of light and air and the other influences of the 'weather.' This process follows an alkaline treatment of the cloth, whether in the earlier or later stages, when the cloth is in the most favourable condition for the action of the atmospheric oxygen. The linen is also treated by a special process of mechanical rubbing with a strong soap solution.

The complications of the process are such that the full linen bleach takes from three to six weeks to accomplish. They are due to the highly resistant character of the cuticular tissues and by-products which are associated with these tissues, or formed during the process of breaking them down; and, in lesser degree, to the wood residues. Both of these have to be entirely eliminated without injury to the cellulose.

The process is therefore a complete illustration of the general chemistry of the compound celluloses, and the order of their resistance to hydrolysis and oxidation, i.e. to the chief destructive influences of the natural world.

It cannot be said that the process has been subjected to exhaustive chemical investigation, such as would reveal the steps by which the various non-cellulose impurities are broken down. From the more theoretical account of these constituents in the earlier sections of this book we may, however, form a tolerably correct estimate of the progress of the breaking-down process. But at the same time a full investigation by chemical and microscopic methods is much more to be desired, and could not fail to throw considerable light upon the important industrial problems involved.

DYEING AND PRINTING PROCESSES.—It appears, *à priori*, that these processes of colouring the textile fibres are the result of interaction of colouring matter and fibre-substance as a definitely molecular phenomenon; and the progress of investigation is confirming this view more and more. At this stage, however, the 'theory of dyeing' is still the subject of active controversy, and a decisive statement must therefore be avoided. The discussion ranges itself round the two opposed views of dyeing: (1) as a *mechanical*, and (2) as a *chemical* process. At the present time, however, these terms have lost much of the significance attached to them in the early days of the controversy. In those days 'solution' itself was regarded as a 'mechanical' or 'physical,' in contradistinction to a 'chemical,' process. As, however, the 'constants of solution,' i.e. the properties of bodies in solution, are now definitely correlated with molecular weight, the distinctions obviously vanish in this case, and the corresponding terms are absorbed in that of more comprehensive significance—viz. 'molecular.' So also it may fairly be stated in connection with the phenomena of dyeing. If solution is defined as the homogeneous distribution of one substance through the mass of another regarded as the solvent, the dyeing *process* is a special case of transference of a body from one solvent to another, and a dyed fibre is a *solid solution* of the colouring matter in the fibre-substance. The conditions determining the transfer, in the process, from water to fibre-substance are certainly complex : they depend (1) upon the constitutional relationships of fibre-substance and colouring matter ; (2) upon osmosis and all those conditions by which it is influenced.

In regard to the first and chief factor, a very superficial view of dyeing processes points to the important influence of the chemical properties of the fibre-substance. But, in extending this view to a detailed discussion, we are met at once by the great disparity between these two groups of carbon compounds, i.e. fibre-substances and colouring matters, in their relationship to the science. The latter are, as a class, bodies of the most definitely ascertained constitution, and are synthesised, in many cases, by 'quantitative' reactions from their constituent groups ; whereas the constitution of the former is still highly problematical in every direction. A comprehensive view of dyeing phenomena is necessarily, therefore, deferred until the latter group shall have been more fully investigated. At the same time, we have positive knowledge of the reactive groups of the fibre-substances, sufficient to indicate the part which they play in dyeing phenomena ; and these reactions have already been discussed, in the case of the celluloses, as a species of double-salt formation. On the more general view of dyeing, this is in fact a well-grounded hypothesis, viz. that as the colouring matters available for dyeing show invariably a 'saline' constitution, and the formation of 'lakes' with inorganic bodies is due to reaction with salt-forming groups—as also the fibre-substances in reaction show a similar differentiation into acid and basic groups—the interaction of the two groups of compounds in the dyeing process is, on the more general view, a special case of double-salt formation. But even should this hypothesis be found to afford a consistent generalisation of the whole range of dyeing phenomena, it carries us only a certain length as a theory of dyeing. We have next to deal with the selective relationships of the two groups of carbon compounds, i.e. the particular 'colouring affinities' of the soluble colouring matters or dye-stuffs. Speaking generally, for instance, the celluloses are resistant to such solutions; the number of dye-stuffs giving a direct dye on cotton is extremely limited. In striking contrast to the celluloses, on the other hand, the lignocelluloses are distinguished by 'cosmopolitan' relationships, resembling the animal fibres wool and silk, in being dyed directly with a wide and varied range of colouring matters. This at once suggests that the essential factors of the dyeing process are molecular and constitutional, i.e. chemical, in the narrow sense of the term, rather than structural; and this conclusion is strongly emphasised by everything which has preceded this discussion in regard to the constitution of these typical groups of fibre-constituents. Further, by chemical modification of the celluloses, their dyeing capabilities are considerably modified; thus the oxycelluloses were shown by Witz to exhibit not merely an increased attraction for colouring matters of the 'basic' class, but a diminished attraction for those of the class more acid in character and generally requiring to be dyed with mordants. Of these two groups the following were cited by Witz as typical. The oxycelluloses show

An increased attraction for	A diminished attraction for
Methylene blue	Diphenylamine blue, sulphuric acid
Hofmann violet	
Malachite green	Induline blue, sulphuric acid
Safranine red	Indigo sulphonate
Fuchsine red	Tropæoline orange
Bismarck brown	Eosine red

in comparison with the cellulose. (See Bull. Soc. Ind. Rouen, [10] 5, 416; [11] 2, 169; Dingl. J. 250, 271; 259, 97; J. Soc. Chem. Ind. 1884.)

Here also structural factors are eliminated, and the variables are again constitutional.

Selective attractions of more narrowly specific character are exhibited, on the other hand, by both the celluloses and lignocelluloses, of which typical instances may be discussed.

Thus, in the case of the celluloses, modern discovery has added to the coal-tar dyes a number of compounds which dye cotton directly to full shades, and are therefore known as cotton colours. Although, however, these are synthetic products, and therefore bodies of known constitution, no general constitutional relationship of these compounds has yet been established such as to account for their 'specific affinities' to the celluloses. This, of course, complicates the phenomena, and shows that other factors, in addition to those of constitution as ordinarily understood, contribute to the result. Of such we may instance as probably operative the molecular condition of the colouring matter *in aqueous solution*.

Of all the colouring matters having this particular relationship to the celluloses, the most noteworthy is the dye-stuff known by the trivial name 'primuline,' a complicated colour-base derived from thiotoluidine. The sulphonic acid of this highly 'condensed' product combines freely with cellulose when the latter is treated with its dilute aqueous solution as in ordinary dyeing process. The combination is of so stable a nature that the base may be diazotised upon the fibre without loss, and then may be further synthesised with chromogenic phenols and bases to form a range of dyes of varying shades. Such 'ingrain' colours constitute an important theoretical and practical advance, and their production by synthetical processes upon the cellulose itself is a further proof that the bond of union of dyestuff to fibre-substance is 'chemical' as ordinarily understood.

Another application of these peculiar relationships of dyestuff to fibre results from the observation that the diazoprimuline upon the cellulose is in a highly photo-sensitive condition, a brief exposure to sunlight sufficing to decompose it with evolution of (gaseous) nitrogen. From this observation has resulted the diazotype process of 'positive' photographic printing (Green, Cross and Bevan, Berl. Ber. 23, 3131).

The important feature of this process, from the point of view of the present discussion, is the sensitiveness of the diazo derivative when prepared upon the cellulose basis, compared with its relative stability in the free state. The most reasonable explanation of this increased sensitiveness appears to be that the product exists in the cellulose in a condition of solution-dissociation, a *solid solution* of the product in the colloid cellulose having the essential characteristics of solutions in liquid solvents. According to this view, the diazoprimuline, being molecularly disaggregated, is in a more 'responsive' condition to the decomposing action of the light-energy; and hence the decomposition. It is no purpose of this discussion, however, to advocate any particular views, but merely to introduce the various aspects from which this in many respects unique dyeing process of the celluloses may be regarded, and to point out that judgment as to the underlying causes must for the present continue to be suspended.

The lignocelluloses afford a still more characteristic dyeing reaction in their property of taking up the blue cyanides from solutions of ferric ferricyanide. It is not a question here of a merely superficial oxidation of the fibre-substance by the ferricyanide, and a staining of the fibre with the resulting blue cyanide. From the detailed description previously given (p. 124) it is seen to be a specific reaction between the fibre-substance and the ferricyanide, taking place in altogether unique quantitative proportions. It does not depend upon any *anterior* reduction by the fibre-substance, as it is unaffected by the presence of powerful oxidising agents; nor upon the relationships to the fibre-substance of either ferric oxide or hydroferricyanic acid, since in any other form of combination they exert but slight action. From the evidence, it appears probable that the lignocellulose takes up the ferric ferricyanide as a whole, in the first instance—such combination having rather the features of a 'physical' reaction—and then redistributes its constituent groups in such a way that the ferric oxide is deoxidised with formation of the blue ferroso-ferric cyanide. In this second effect the constitution of the characteristic groups of the lignocellulose is the active cause.

These two reactions or groups of dyeing phenomena have been instanced, not only because they are of critical and unique value as test-problems for any theory of dyeing, but as further illustrating the varied aspects of the subject of cellulose chemistry. With progress in the theory of dyeing, it is highly probable that the effects themselves may come to be available as criteria of constitution of the fibre-substances; in the mean time it is equally probable that further elucidation of these problems in other directions may contribute materially to the establishment of a theory more generally acceptable than the much controverted views at present held.

In the processes of printing the vegetable textile fabrics the same general considerations obtain. The treatments are, however, much more diversified; and their scientific basis, so far as regards the chemical function of the fibre-substance as an active cause, is even less elucidated than in the more simple operations of dyeing. In the absence of any specific contributions of investigators, no attempt can be made to deal with so wide a range of effects. With a wider knowledge of the chemical functions of the constituent groups of the fibre-substances, it will be easy to devise critical experiments in solution of the very various problems presented.

The industrial uses of the celluloses and compound celluloses are of wide and varied range. They depend, of course, largely upon the external and physical properties of the natural products: but if less obviously, certainly in a not less important degree upon the special chemistry of these substances. Their industrial value again depends upon the conditions of supply, the agricultural questions of yield, and the economic questions of production and preparation in a fit state for the further manufacturing operations by which they are finally shaped for use.

In the province of textile fibres this threefold qualification constitutes an effectual limitation of the number available, and the numerous abortive attempts to exploit others of the endless variety of vegetable fibres have invariably followed from neglect of one or other of the essential conditions of qualification. These qualifications are in effect the *constants* of the fibres, all expressible in numbers, the results of measurements or observations of quantitative relationships. Thus, to select in illustration the flax fibre, the following are the 'constants' which mainly determine its value:

Agricultural (constants of raw material) . . . } Yield of 'straw' per acre.
Yield of fibre on 'straw.'

Morphological or structural (physical constants of fibres) } Length of ultimate fibre.

Chemical . . . . { Proportion of cellulose and resistance of cellulose to hydrolysis and oxidation.

There are many considerations of subsidiary importance: thus, on the agricultural side, the habit of the plant and cost of cultivation; on the mechanical or structural side, the separation of the fibres from the stem, the uniformity, fineness, and divisibility of the fibre-bundles; and on the chemical side, the relationship of the cellulose to the non-cellulose constituents both adventitious (wood and cuticle) and essential (the pectic constituents of the fibre proper). A careful consideration of these quantities or properties as factors of value will almost tempt the reader, if of a mathematical turn of mind, to propose a numerical expression of value somewhat as follows:

Taking V = value (in the sense of utility),
Y = yield of fibre per acre,
L = length of ultimate fibre,
P = percentage of cellulose in fibre,
then $V = c.\ YLP$ (c being a constant).

The factors Y, L, P would require to be qualified by the introduction of the subsidiary factors; and although these are not expressible in so definite a form, they can be brought to a sufficiently exact approximation. It is not the purpose of this inquiry, however, to attempt a complicated special discussion involving considerations outside our general plan of treatment. With this general suggestion of the relationships of our subject, taken as a whole, to industry, we revert to the consideration of the purely chemical problems presented by the celluloses and allied compounds in use. These problems are in effect those of destruction and disintegration.

Of the textile fibres cotton and flax are by far the most important, and the position which they occupy is very largely determined by the properties of their cellulose basis. This cellulose is amongst C.H.O compounds very much what silver and gold are amongst the metals, manifesting, that is, a high degree of resistance to the chief disintegrating agencies of the natural world—oxygen and water. Both fibres have been used from the remotest antiquity, though the manufacture of cotton textiles in Europe is of quite modern growth. At the time of its introduction it was used for padding and filling purposes and for manufacture into paper. The spinning of the short staple fibre into yarn is an art borrowed from the East, where it has been practised from the remotest antiquity.

Of both cotton and flax, however, we have sufficient record —in the substantial form of manufactured products—to be able to pronounce them for practical purposes indestructible save by the mechanical agencies of wear and tear. In ordinary use, however, they require periodical cleansing; and the severe treatments of the laundry, chemical and mechanical, lead to more or less rapid disintegration. Very little attention is paid to this industry from the chemical point of view, of which the chief regulating principles are those of economic and rapid handling. Occupying as it does a somewhat 'inferior' position in human affairs, it appears to be beneath the notice of technologists. The result is unfortunate, as the very common experience of the household will testify.

The cleansing of vegetable textiles by alkaline solutions, wherever and however practised, is a chemical process; and it is high time that laundry work, conducted as it now is upon the scale of an enormous special industry, should be more consistently organised as a chemical industry. Great progress in this direction would be made by modelling the procedure of the laundry upon the general principles of treatment of these textiles in the manufacturing industries; i.e. in the case of cotton and linen goods the lines of treatment should be generally *similar* to those of bleaching and finishing—though, of course, differing considerably in degree. As a matter of experience, the chemical disintegration of these textiles in the course of laundrying is considerable, chiefly through ignorance or neglect of the chemical properties of the celluloses on the one hand, and the cleansing agents employed on the other. This, again, is a subject opened up in definite directions of inquiry by the matter of this treatise, and it is to be hoped that the chemical history of a shirt or tablecloth may come to be written at no distant period, and with special attention to those conditions which make for longevity.

Of the uses of vegetable textiles in their unbleached or partially bleached conditions there is little to be said from the chemical side. It should be remembered that half-bleaching treatments are fraught with some danger, owing to the chemical changes (oxidation or chlorination) in the residual non-cellulose constituents; and to minimise these dangers a final treatment with sulphite or bisulphite of soda is to be recommended. The authors have in mind not only the facts in connection with jute bleaching mentioned on p. 286, but have been called in to adjudicate upon damages occurring in the bleaching 'out' of flax goods woven with creamed or half-bleached yarns. These have been frequently found to retain substantial quantities of 'chlorine' (bleaching powder), and it is quite remarkable the length of time of persistence of these residues of hypochlorites in contact with flax goods. Their presence must involve a gradual oxidation of the entire fibre-substance, which together with the acidity of the oxidised non-cellulose effects a steady disintegration of the fabric. So long indeed as goods are treated altogether without reference to the molecular results of the treatments, sound practice is the result of tradition and correct intuitions, and the chances are far too numerous in favour of malpractice. If at any time there should be an extensive exposure of the secrets of the 'damage room,' it might occasion wonder that a stronger case should not have been made out for the scientific regulation of these industries.

The second great branch of the cellulose industry is that of paper. Here also we meet with a large proportion of fabrics composed of unbleached or partially bleached materials, in reference to which there is little to be said from the point of view of their chemistry. They are used for 'inferior' purposes, such as wrappings; they serve their purpose, and there are no problems of especial import presented by the chemical history of the fibres in this particular form.

But it is otherwise with papers used for writing and printing. In this category permanence is a first desideratum. Books and records have more than a passing value, and it is essential that they should be committed to pages suitably resistant both to chemical and mechanical wear and tear. On the other hand, we may safely affirm that there is no public opinion in this country upon this important subject. Where preferences for high-class papers exist they are based rather upon æsthetic and other recondite considerations than upon any judgment as to composition and the relation of their constituents to the destructive agencies of the natural world. On this basis white papers admit of a very simple classification into three main groups: (A) those composed of the normal and resistant celluloses only—e.g. cotton, linen; (B) those composed of celluloses containing oxidised groups or oxycelluloses—e.g. wood-cellulose, esparto and straw celluloses; (C) those containing, in admixture with the above, ground wood or mechanical wood pulps (lignocellulose), many of which are sufficiently 'white' as not to prejudice a paper from the point of view of colour.

Of the above, Class A stands beyond criticism. From the discussion of the chemistry of the celluloses it is evident that they fulfil all the requirements of inertness, and this may be taken as a confirmation of the extensive experience which we have of the lasting properties of the celluloses. Throughout the middle ages these fibres were the staple raw materials for production of papers, and in books that have come down to us from these times there is sufficient evidence of resistance to the natural processes of disintegration.

Fibres of Class B have been introduced in response to the enormously increased consumption of paper in this century, and it becomes important to consider how far they fail, or may on chemical evidence be predicted to fail, in regard to the properties which distinguish the former class. It is evident that chemically they are of totally different constitution, esparto and straw diverging from the normal type much more considerably than the wood-celluloses. It is a matter of observation that all papers containing these celluloses are liable to discolouration under the ordinary conditions of wear and tear. Chemists will have made the further observation that in the atmosphere of the laboratory, reference books, or rather the paper upon which they are printed, are liable to peculiar discolourations. Thus, in laboratories where coal-tar products are handled it is a frequent experience that our journals change from white to bright pink, and even where there is no direct contact with the atmosphere of the laboratory it is common to see the pages change to various shades of brown. This browning can be produced in a very short time by exposure to the heat of the water-oven, and it has also been shown that under these conditions the fibre undergoes oxidation which is sufficiently marked to be measured by an increase of yield of furfural on boiling with hydrochloric acid. It is clear, therefore, that these reactive oxycelluloses are inferior in an important chemical sense, and their use in books is open to the very obvious objection that the books are more perishable. Of course, it is perfectly true that a large amount of literature is of the ephemeral kind, and in this province such questions as we have raised do not enter; on the contrary, paper being very much cheapened by the use of these celluloses, a great advantage is gained. It must be insisted upon, however, that authors and publishers should have a definite judgment as to the papers to which they commit their productions, and it would be of the greatest utility to exhaustively investigate these particular celluloses from the point of view of their resistance to the natural processes of decay.

CLASS C.—The presence of lignocellulose is a more extreme departure from the sound basis of composition represented by Class A. The lignocelluloses are not only more generally reactive than the celluloses of Class B, but are easily attacked by atmospheric oxygen (see p. 174). Added to these chemical defects they are inferior in the mechanical properties which contribute to the strength of the sheet of paper, and therefore papers of this class are only permissible where lasting properties are a question of no moment whatever.

In addition to these questions of the composition of the fibres or pulps, the practice of loading papers with china-clay, sulphate of calcium, and so forth, is also another of the causes which lead to disintegration of modern papers as compared with those of former days. There is, of course, the other side to this question, the addition of these mineral diluents having certain positive advantages not to be overlooked. The danger of any practices of this kind only enters when they are not measured at their proper utility. Paper is largely 'taken for granted' by consumers. In a great many, perhaps the majority of cases this unenquiring consumption is not attended with any serious consequences; but, on the other hand, it is quite obvious that it is attended with dangers of a very grave character, when we are dealing with records of value for all time. This, of course, is largely a question for posterity, to whom we are handing down a literature produced upon grounds for the most part of mere commercial expediency.

It is high time, as we have said before, that a public opinion should be formed upon this subject, and it can only be formed upon a recognised classification of papers, based upon their chemical and mechanical 'constants,' which are determinable by laboratory investigation. In Germany considerable progress has been made in the fixing of standards of quality and securing their adoption by the trade. This classification by fixed standards has been systematically worked out in the Government Testing Station at Charlottenburg, and the records of the institution contain a number of important monographs upon the various factors of quality of papers. As these, however, contain no very direct contributions to the chemistry of cellulose, we have only to call attention to the general result of the investigations. In our own country the character of the paper trade differs in many respects from that of the Continent; and this would necessitate a special classification and series of standards. So far, however, as this classification is based upon differences of chemical composition the lines of demarcation are simple and sharp, and the general recognition of these will initiate a movement in the direction of *specific* uses of papers according to their qualities and properties.

Outside the province of textiles and papers there are many other uses of cellulose of great industrial importance, many of which have been dealt with incidentally in the foregoing pages. The nitrates of cellulose are the basis of manufactures which have been developed within our own period of history. They are used on the one hand as a plastic and constructive material, on the other as an explosive and destructive agent; these uses affording remarkable illustrations of the chemical and physical properties of cellulose. In regard to the former, the use of the nitrated compounds of cellulose is open to the very obvious objection of high inflammability. The combined nitric acid is in fact a necessary evil; and from what we now know of cellulose in aqueous solution as thiocarbonate (p. 25), its 'gratuitous' character becomes still more prominent. The nitric groups are merely a factor of a particular process of solution of cellulose; they do not modify in any essential respect the properties of the parent molecule, but render these available by bringing the cellulose into a condition of homogeneous solution. Lehner's 'artificial silk' process illustrates these considerations in a very direct way. For the spinning of the thread the solution as nitrate is necessary; but the subsequent process of denitration changes the physical properties of the product in so small a degree as to escape detection otherwise than by the application of special tests. The products known as celluloid, xylonite, &c., are not subjected to any denitration process; but the cellulose products obtainable by means of the cellulose xanthate are so similar to these that the plastic properties of cellulose itself are more than ever apparent as the essential basis of these manufactures. The same facts are illustrated by the acetates of cellulose. When these are prepared under carefully regulated conditions they exhibit the same properties in solution as the nitrates, i.e. high viscosity and coalescence, on evaporation of the solvent, to a homogeneous elastic solid. It is evident, therefore, that the nitrates of cellulose in such uses will be subjected to the ordeal of a severe competition, and in certain directions must be displaced by the parent substance itself or by derivative compounds at present known or yet to be discovered.

The manufacture of explosives composed exclusively or partly of the cellulose nitrates is now an industry of enormous proportions. For many years after the introduction of gun-cotton as an explosive its application was limited by its denomination as a 'high explosive,' i.e. for blasting and similar purposes. The researches of later years have shown that by changing the physical condition of these 'high explosives' their explosive combustion may be brought under perfect control, and they therefore become available as propulsive explosives, i.e. in artillery and small arms. In these directions they are rapidly displacing the charcoal or black powders which have done so much service to the human race in the past centuries! A special advantage of these nitrocellulose powders from the military point of view is that, owing to their perfect combustion to gaseous products, their explosion is a 'smokeless' one: hence their general and popular designation. The basis of these 'powders' is a mixture of nitroglycerin and 'nitrocellulose.' The nitrates of cellulose are gelatinised by nitroglycerin, and by varying the proportions homogeneous plastic mixtures of varying consistency are obtained. With small proportions of the cellulose compounds, 7-8 p.ct., a gelatinous mass is obtained, known industrially as 'Blasting Gelatine.' With lower proportions, gradations of consistency are obtained in the mixture which is the basis of explosives of the 'Gelignite' class. With the cellulose nitrates increased to 40-50 p.ct. a semi-solid product is obtained, which is worked up into threads or ribands and constitutes the military smokeless powders ('ballistite,' 'cordite,' &c.). The product resulting from the mixture of these two 'high explosives' burns quietly when ignited, and, burning from the *surface*, the combustion is perfectly under control, and can be easily regulated to avoid detonation. A second class of 'powders' is made by mixing the nitrocellulose and a certain proportion of barium nitrate with a smaller proportion of camphor or nitrobenzene to allow of their being worked up to a suitable form. Such are the 'E.C.,' 'S.S.,' and other 'sporting' powders. In many of the latter the nitrocelluloses employed are prepared by nitrating the celluloses of Class B (*supra*) isolated by the processes of the papermaker; in some cases also nitrated lignocelluloses are employed.

These industries are in a highly developed condition, the

manufactures being carried on with the greatest precision, on the basis of an extensive empirical knowledge of the properties of the products. It must be admitted, however, that, in the absence of any precise knowledge or even accepted theories of the constitution of the cellulose nitrates, there remains a vista of progress to be opened out by the solution or partial solution of this important problem.

So, in fact, it may be said, generally and in conclusion, of the industrial uses and treatments of the celluloses. All of great and some of the greatest importance in human affairs, and all highly developed upon an extremely slender foundation of exact knowledge of the raw materials, it is probably true that the cellulose industries have in many directions attained a position of terminal excellence, measured from the point of view of an empirical technology of the subject-matter. It may be said with greater certainty that future progress will go hand in hand with the progress of scientific investigation.

It is a province of applied chemistry where, as in many others, the distinctions between 'Science' and 'Practice' exist only in the minds of those who grasp neither the one nor the other. Manufacturers and technical men, if they will only take the trouble to inform themselves, must see that an enormous field of natural products and processes about to be explored has a number of industrial prizes and surprises in store; scientific men who have to undertake the pioneering work in this field will find sufficient stimulus to effort in the promise of progressive discovery. It is to be hoped that some suggestions of matter for research will be conveyed in the foregoing brief account of the present position of the chemistry of cellulose.

APPENDIX I

THE illustrations which follow, reproduced from sections of typical raw materials from amongst those dealt with in the preceding pages, are designed to convey an outline view of their general features of structure and arrangement in the plant.

The subjoined scheme of classification of fibrous raw materials is based upon these structural or anatomical features considered as the necessary basis of their varied applications in the arts. The selection of types in illustration has been made in accordance therewith, and as it is a sufficient key to their selection and arrangement it is reproduced (from Indian Fibres, p. 18) without further comment or explanation in detail.

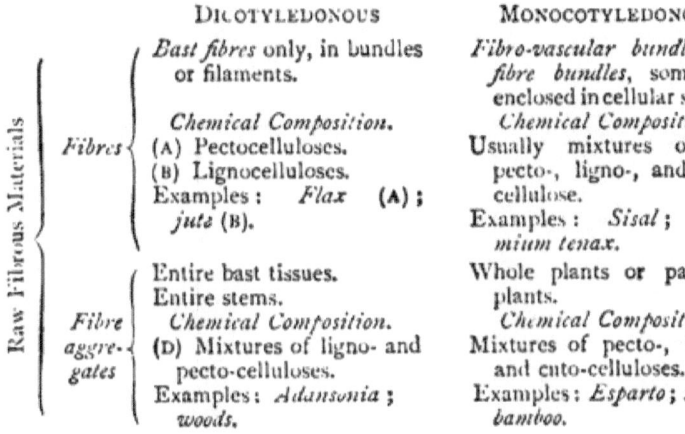

	DICOTYLEDONOUS	MONOCOTYLEDONOUS
Fibres	*Bast fibres* only, in bundles or filaments. *Chemical Composition.* (A) Pectocelluloses. (B) Lignocelluloses. Examples: *Flax* (A); *jute* (B).	*Fibro-vascular bundles* and *fibre bundles*, sometimes enclosed in cellular sheath. *Chemical Composition.* Usually mixtures of (C) pecto-, ligno-, and cuto-cellulose. Examples: *Sisal*; *Phormium tenax*.
Fibre aggregates	Entire bast tissues. Entire stems. *Chemical Composition.* (D) Mixtures of ligno- and pecto-celluloses. Examples: *Adansonia*; *woods*.	Whole plants or parts of plants. *Chemical Composition.* Mixtures of pecto-, ligno-, and cuto-celluloses. Examples: *Esparto*; *straw*; *bamboo*.

It is important to note that the cotton fibre—the chemical prototype of the celluloses—does not fall within the above classification. As a 'seed hair' it stands apart. The cutocelluloses are non-fibrous, and constitute a structural class (E) also outside the above, though occurring in C and D in admixture with the fibrous constituents proper.

PLATE I.

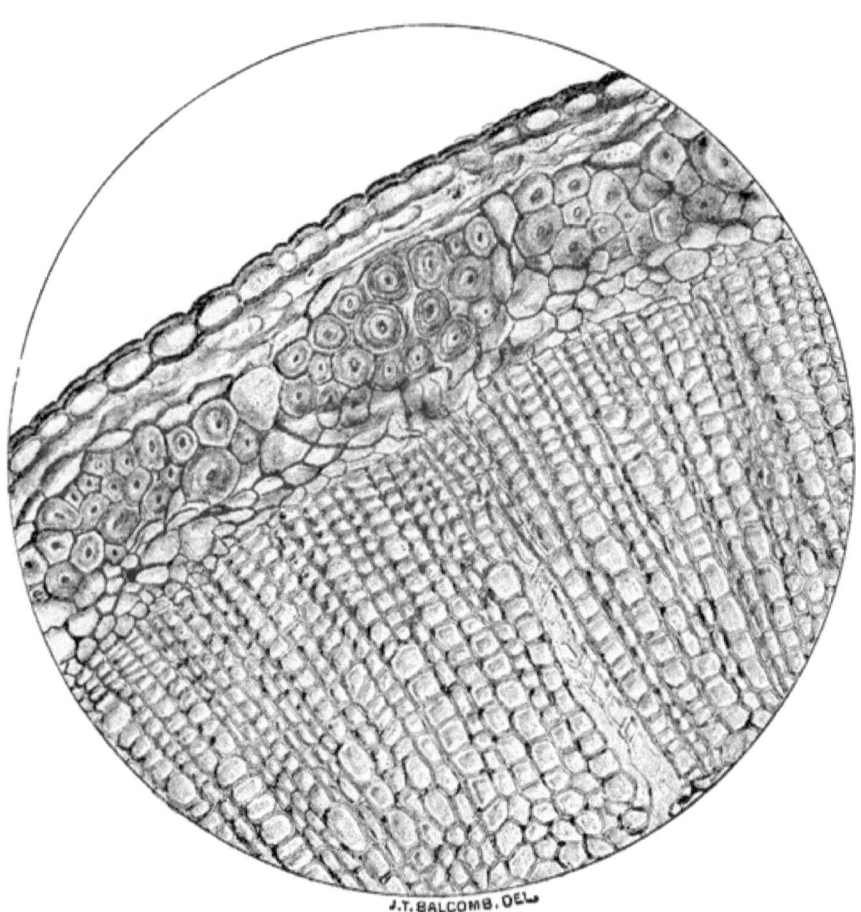

J.T. BALCOMB, DEL.

A.[1] 1. FLAX—*Linum usitatissimum.* × 150.
  Transverse section of stem.
  Beginning at periphery:—
    Layer of cuticular cells.
    Intermediate cortical parenchyma.
    Bast fibres in groups—*flax fibres* proper. Note secondary
      thickening of cell walls.
    Cambium region.
    Wood.

[1] These letters refer to the grouping of the table, page 311.

PLATE II.

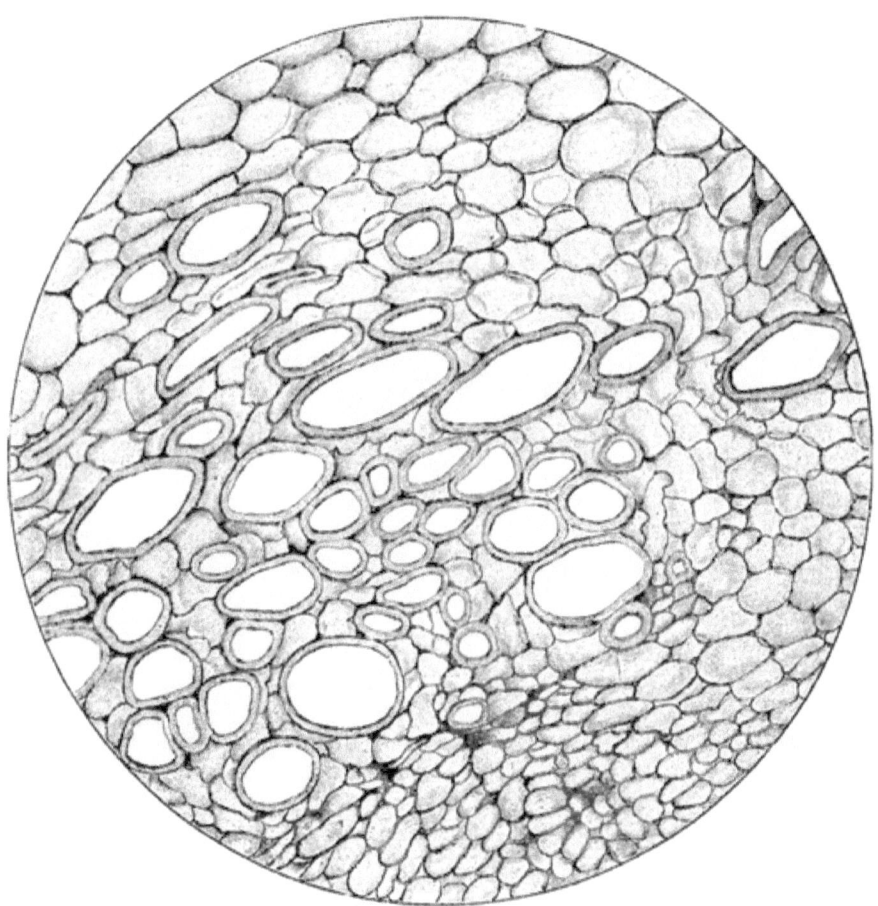

A. 2. RAMIE, RHEA, OR CHINA GRASS—*Bœhmeria nivea*
× 150.
Transverse section of bast region only.
Bast fibres, distinguished by their large area from adjacent tissue.

PLATE III.

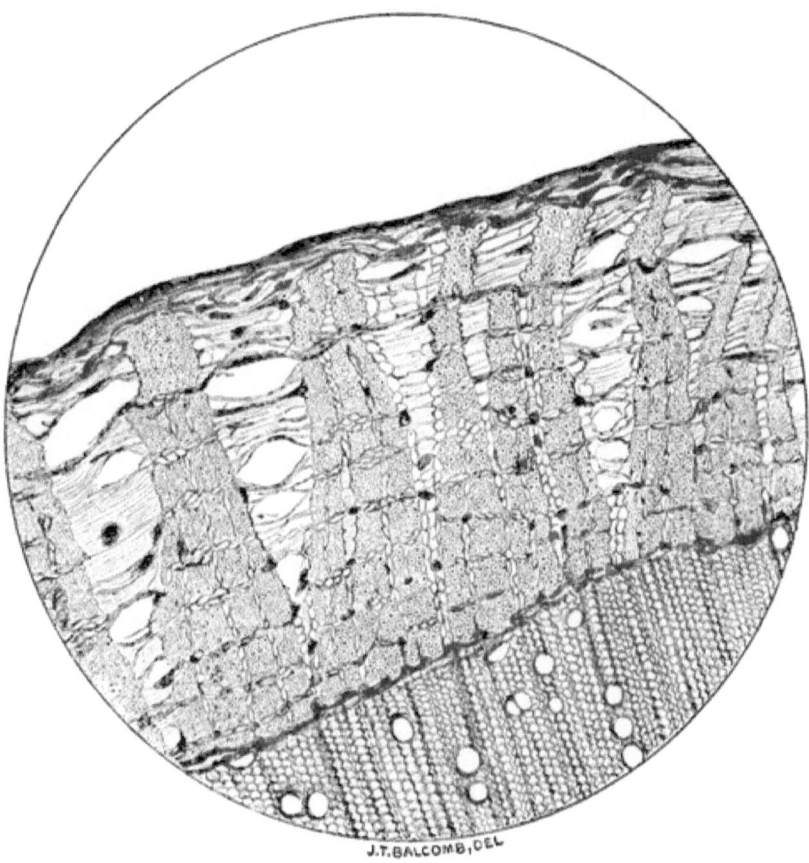

B. 3. JUTE—*Corchorus capsularis.* × 50.
Transverse section of stem.
Wedge-shaped complexes of bast bundles extending from the cambium to cortex.

PLATE IV.

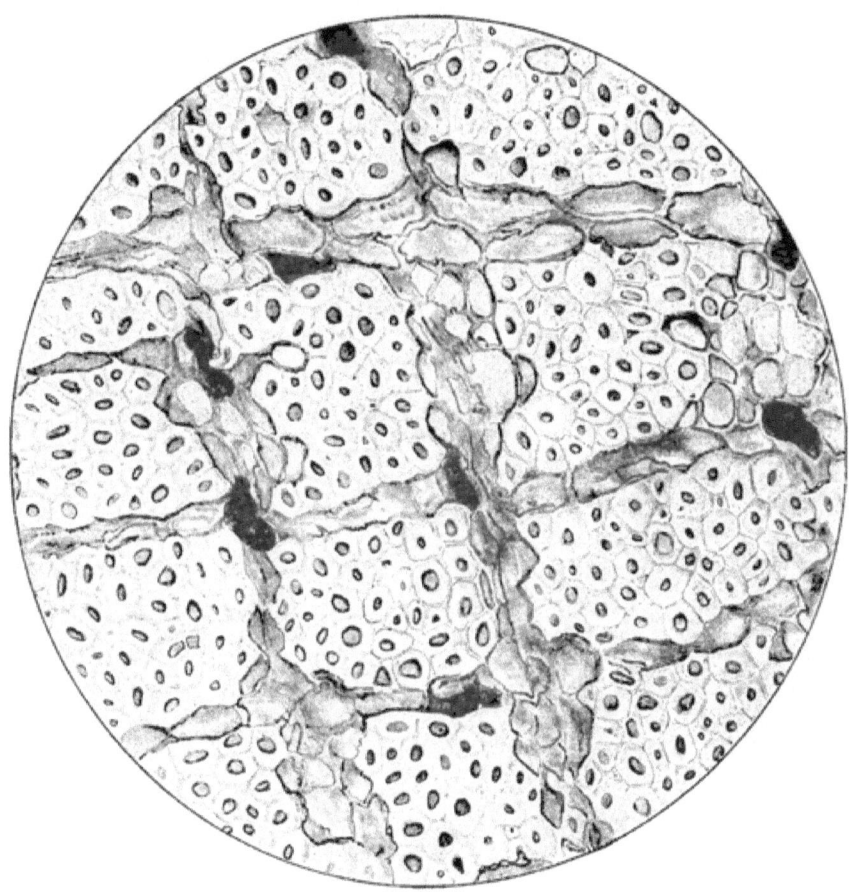

B. 4. JUTE.- *Corchorus capsularis.* × 300.
Transverse section of portion of bast. Showing anatomy of fibrous tissue, form of bast fibres, and thickening of cell walls.

PLATE V.

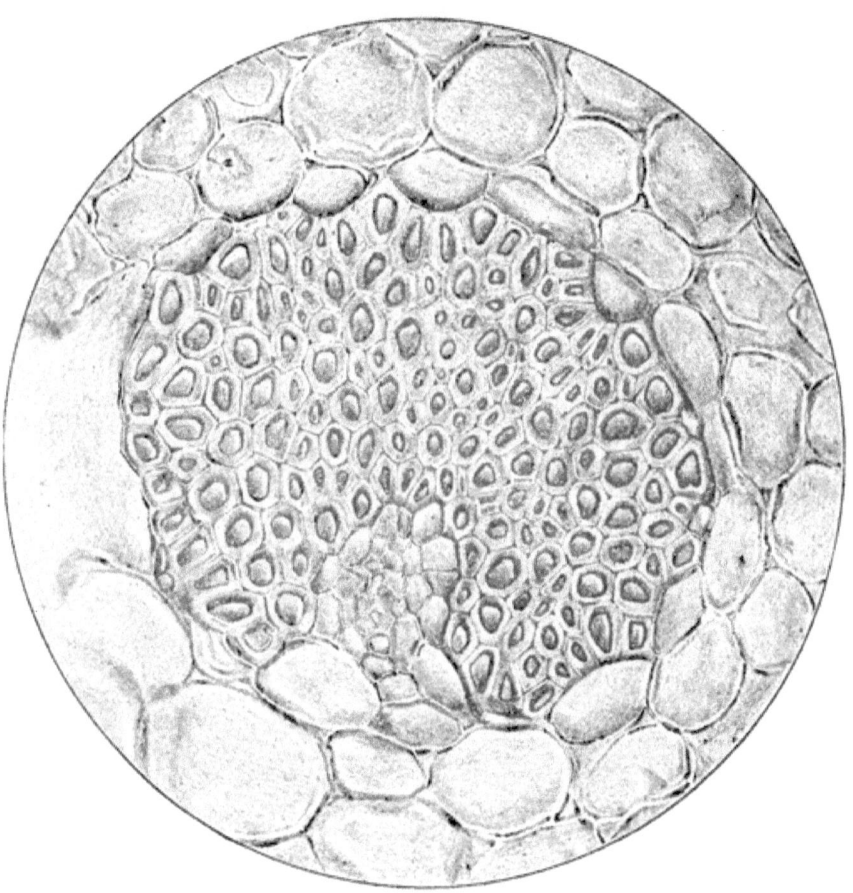

C. 5. SISAL HEMP—*Agave Sisalana*. × 300.
Transverse section of single filament.
Kidney-shaped complex of lignified fibres almost enclosing vessels, the whole surrounded by parenchymatous tissue.

PLATE VI.

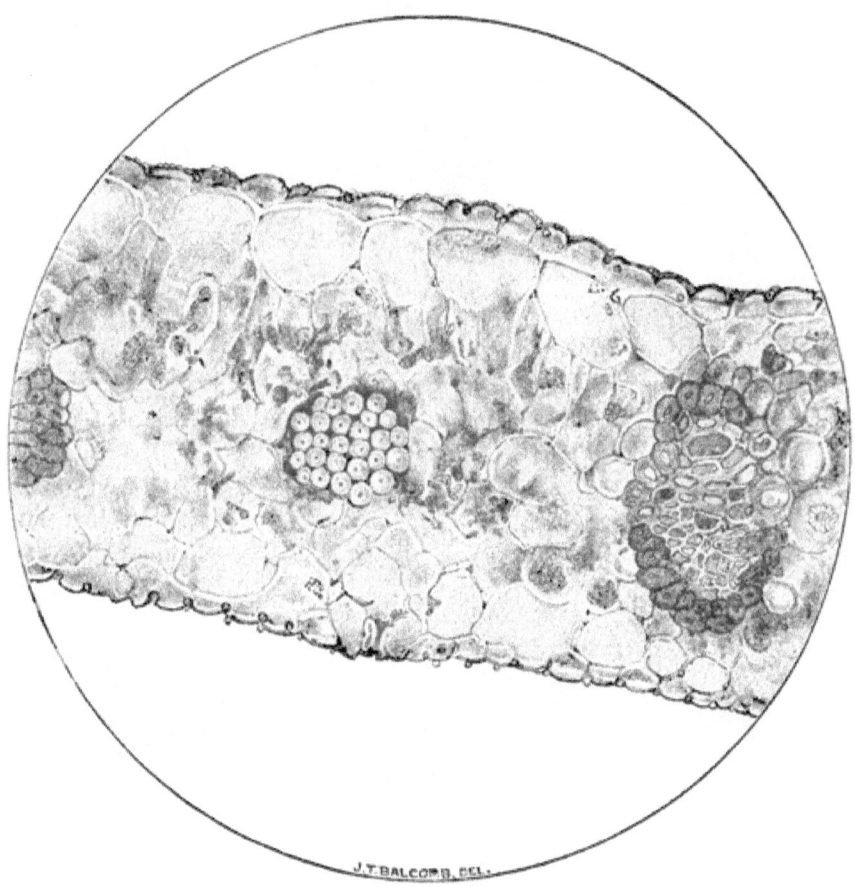

C. 6. OIL PALM LEAF—*Elæis guineensis.*   × 300.
   Transverse section of part of leaf, showing two classes of filaments:—
   1. Large fibro-vascular bundle.
   2. Smaller bundle of thick-walled fibres without vessels.

PLATE VII.

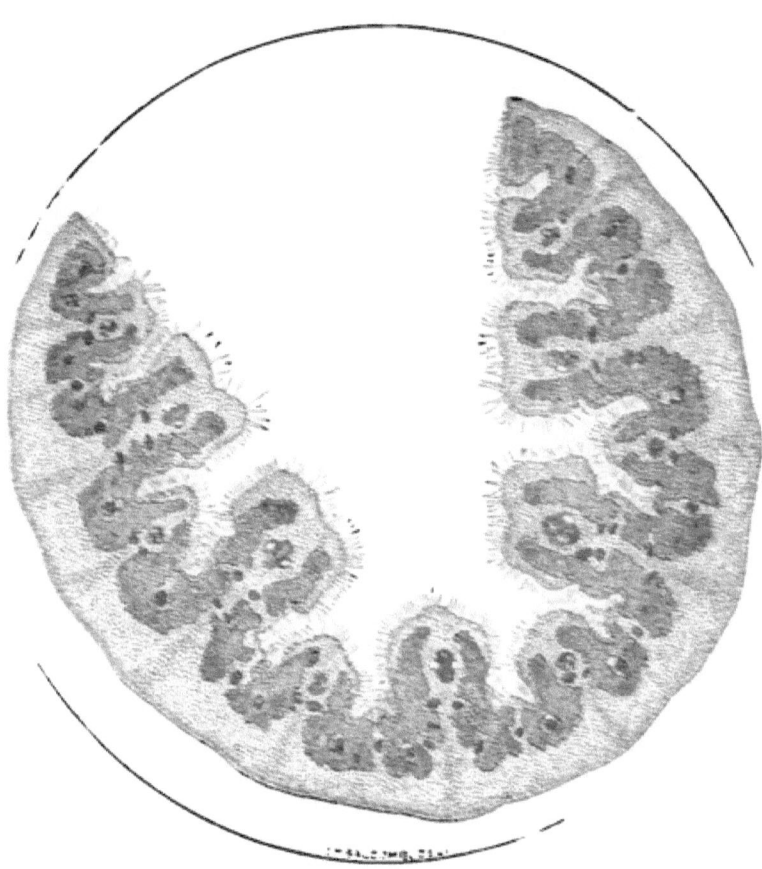

Fig. 7. ESPARTO—*Macrochloa Stipa tenacissima.* × 50.
Transverse section of leaf.
Upper side composed of projecting ribs and deep bays fringed with siliceous hairs. Areas of chlorophyll-bearing parenchyma interspersed with fibro-vascular bundles.
Lower side composed of parenchymatous fibres. In central region of each rib, bands or bridges of thick-walled lignified cells extending from lower to upper epidermis.

PLATE VII.

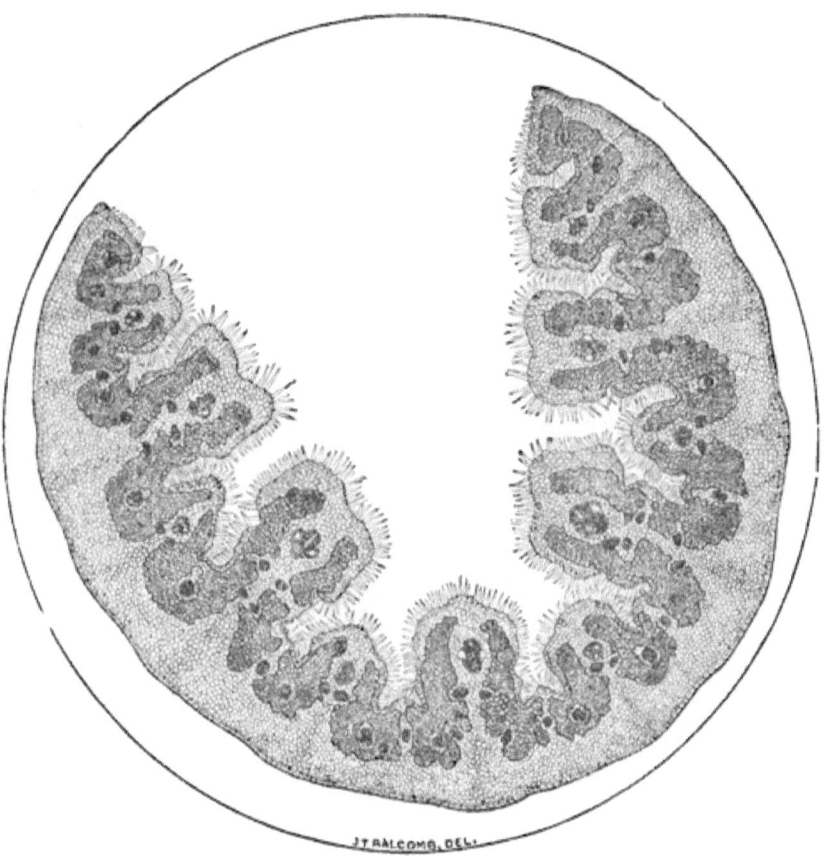

D. 7. ESPARTO—*Macrochloa (Stipa) tenacissima.* × 50.
   Transverse section of leaf.
   Upper side composed of projecting ribs and deep bays fringed with siliceous hairs. Areas of chlorophyll-bearing parenchyma interspersed with fibro-vascular bundles.
   Lower side composed of prosenchymatous fibres. In central region of each rib, bands or bridges of thick-walled lignified cells extending from lower to upper epidermis.

PLATE VIII.

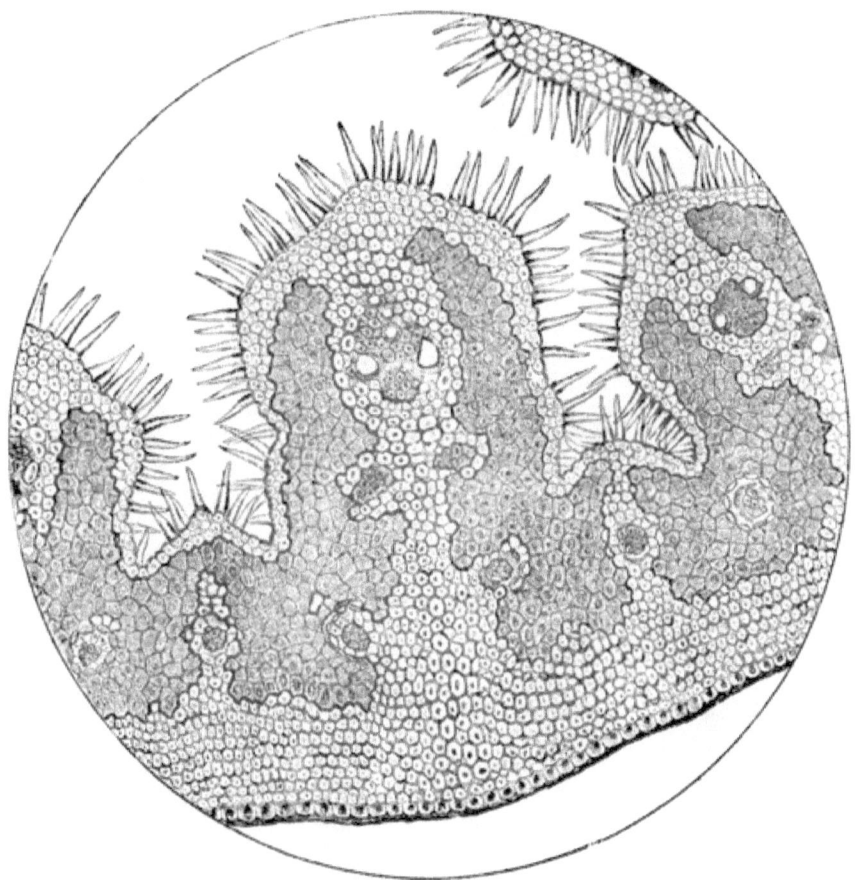

D. 8. ESPARTO—*Macrochloa (Stipa) tenacissima.* × 150.
  Transverse section of central ribs, &c.
  General features of preceding section in greater detail, and showing more clearly the band or bridge of lignified tissue passing from lower epidermis between the chlorophyll areas and surrounding the large fibro-vascular bundle.

PLATE IX.

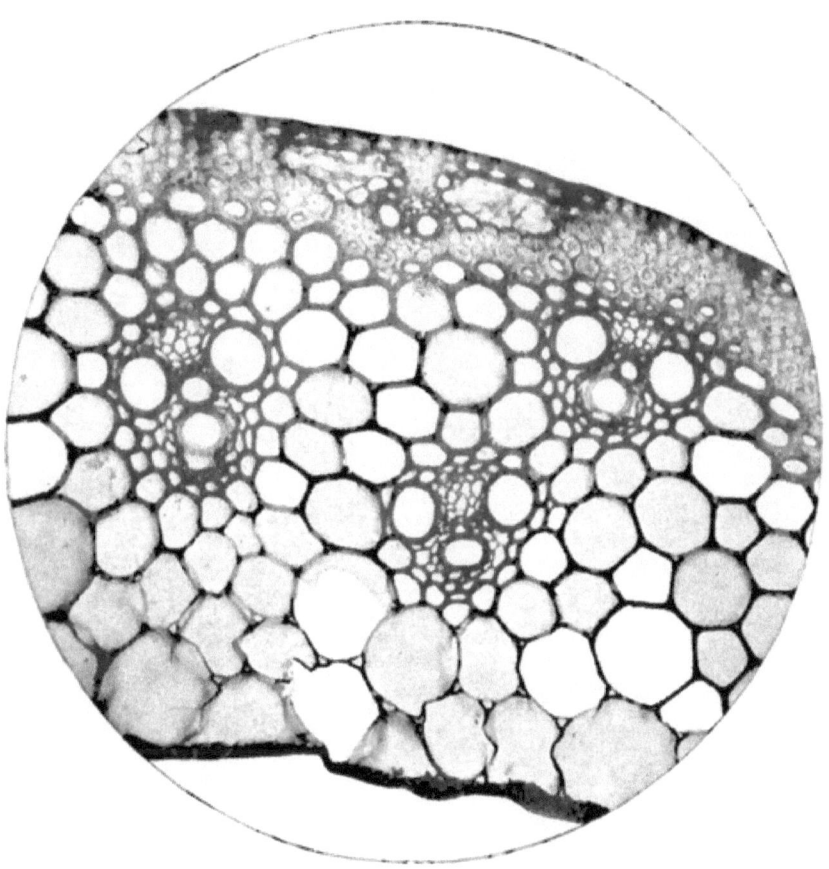

D. 9. Straw (Wheat)—*Triticum vulgare.* × 150.
    Transverse section of stalk.
    Hypodermal layers composed of strongly lignified and thickened fibres with small fibro-vascular bundles.
    Larger f.v.b. disposed through thin-walled parenchyma.

PLATE X.

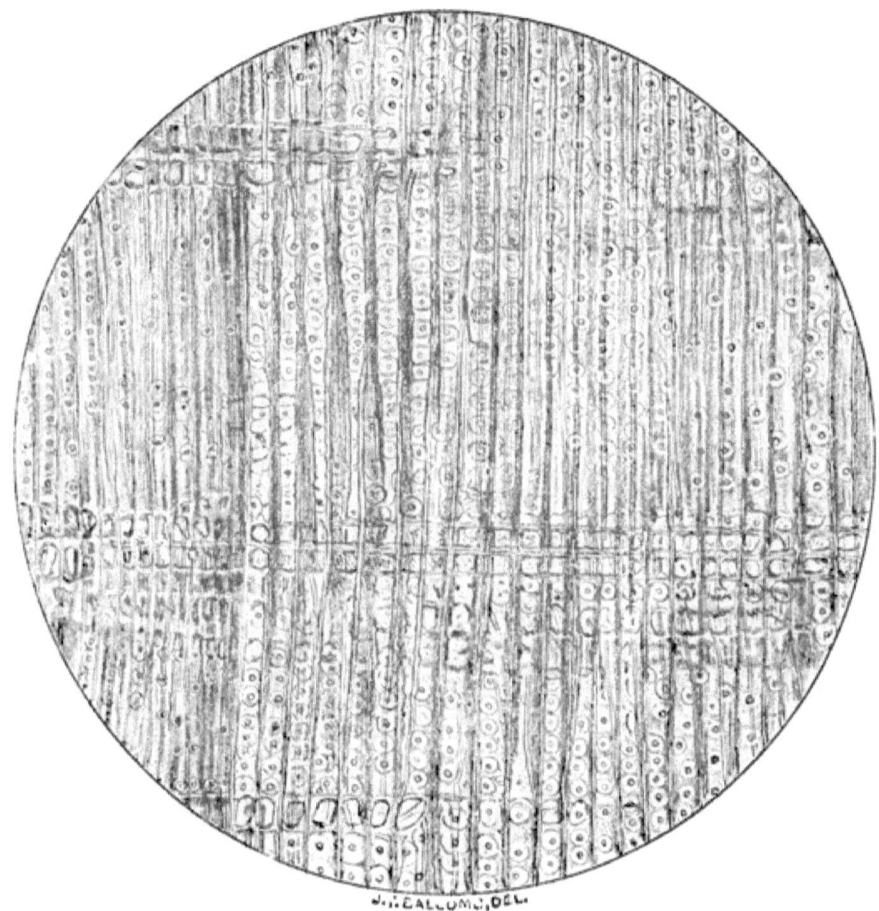

D.   10. WOOD—*Pinus sylvestris.*   × 150.
   Longitudinal section.
   Tissue chiefly composed of the characteristic tracheides with numerous 'bordered pits,' intersected by medullary rays.

PLATE XI.

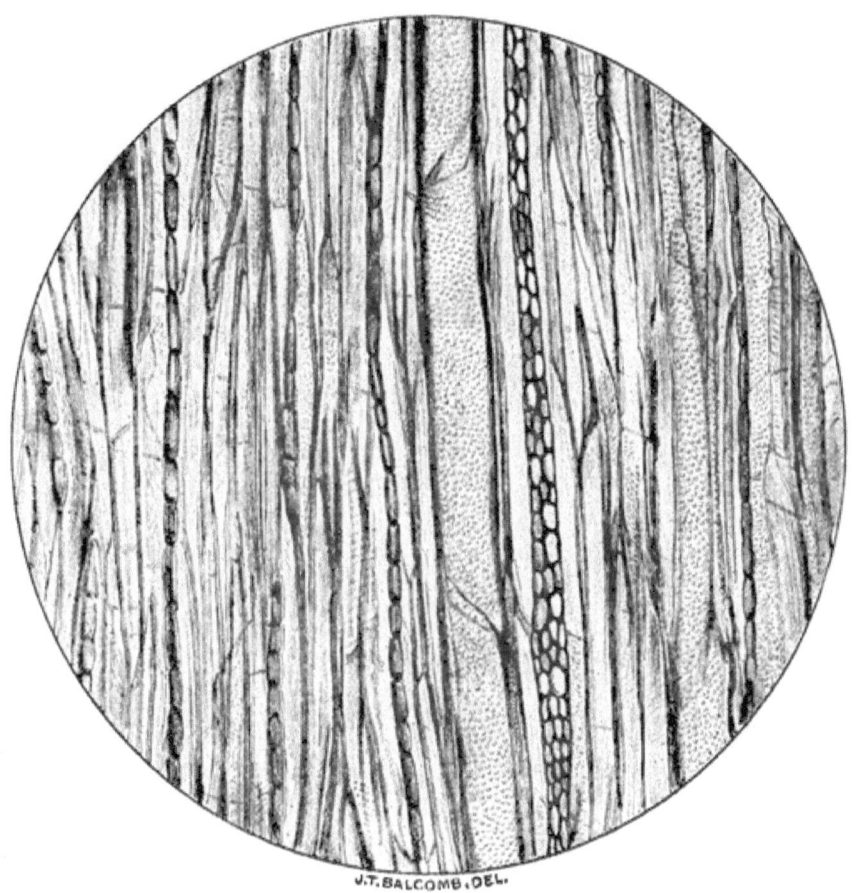

D. 11. Wood—*Tilia grandiflora.* × 150.
    Longitudinal section.
    Wood fibres, woody parenchyma, and large pitted vessels with oblique septa.

PLATE XII.

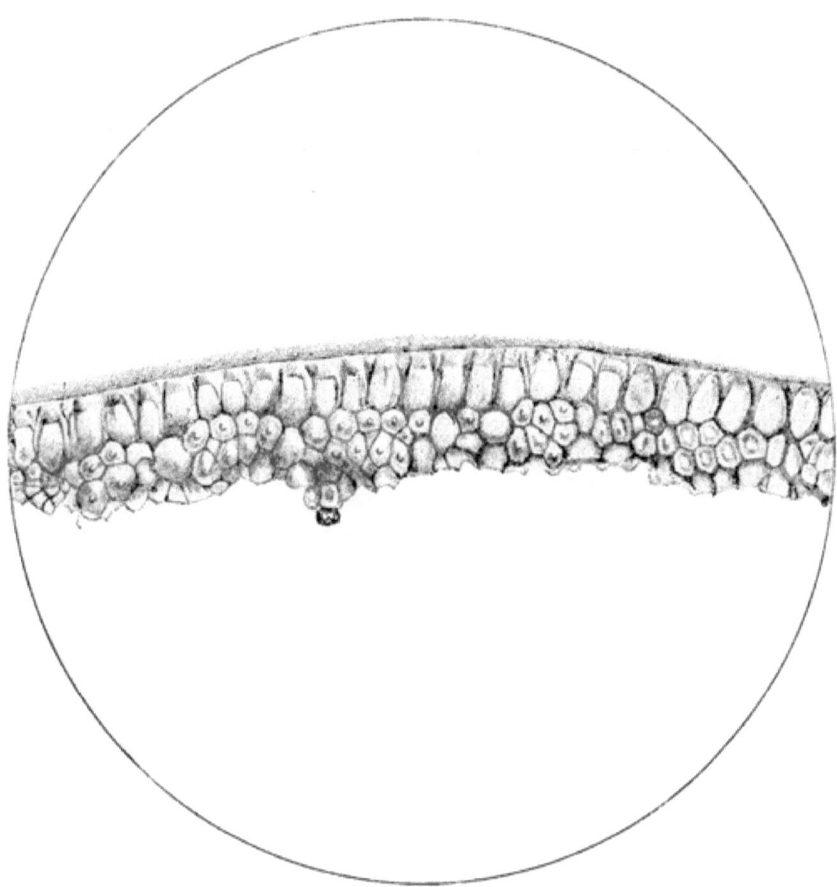

E. 12. RAFFIA—*Raphia Ruffia.* × 300.
   Transverse section of epidermal tissues constituting the commercial fibre.
   Cortex of upper surface, with bundles of hypodermal fibres with strongly thickened walls.

PLATE XIII.

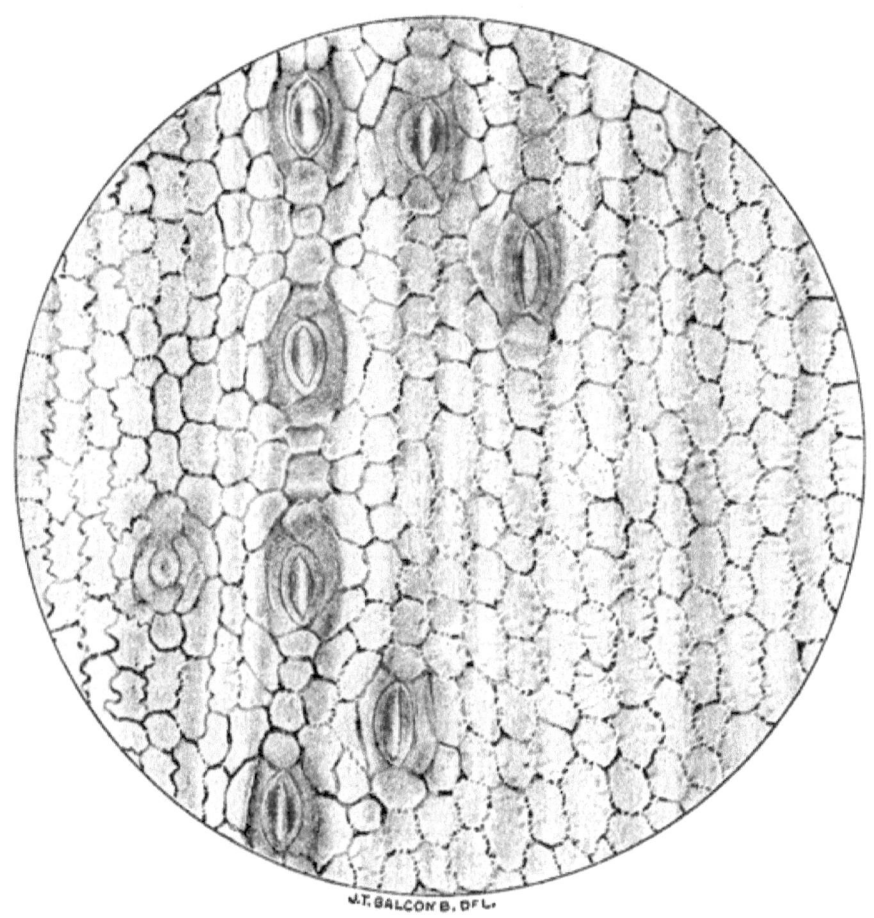

E. 13. RAFFIA *Raphia Ruffia.* × 300.
 Surface view.
 Cortical cells with serrated outline and stomata.

PLATE XIV.

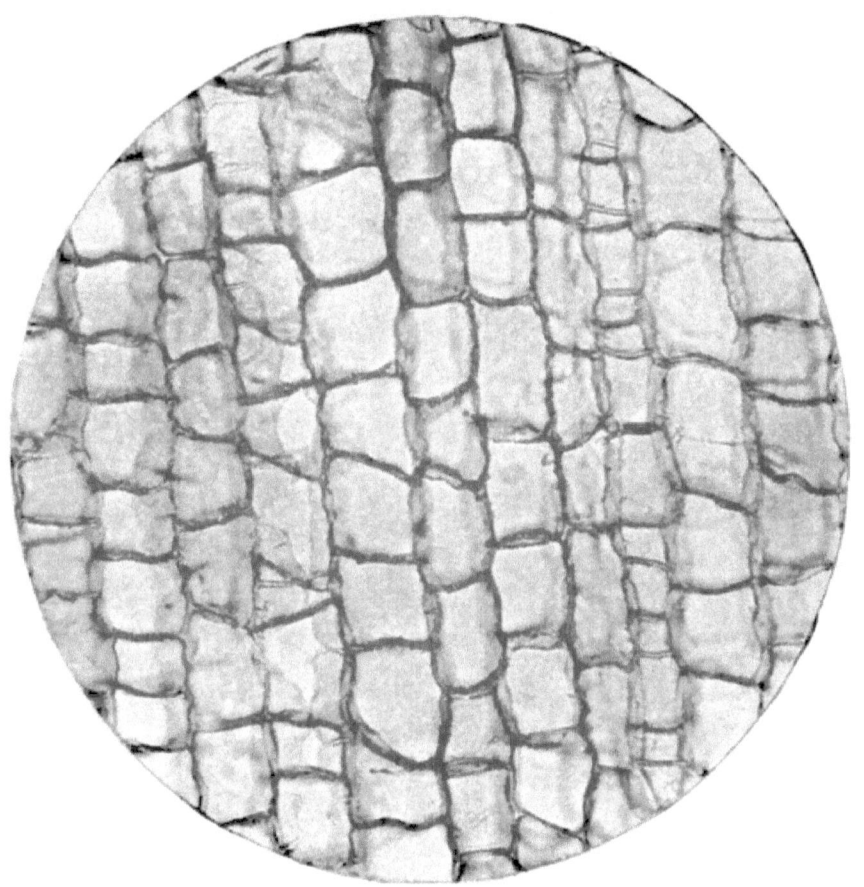

E. 14. BOTTLE CORK—*Quercus suber.* × 300.
  Transverse section.
  Thin-walled cork cells.

APPENDIX II

In the period 1895-1900 succeeding the publication of the first edition of this work, there have appeared a number of contributions to the general chemistry of 'Cellulose,' the more important of which have been recorded and discussed in a volume of 'Researches on Cellulose' by the present authors, published in 1901.

It will be of interest to our readers to follow the main lines of growth of the subject; and we therefore give a brief outline, and in very general terms, of these later developments.

Cellulose.—*Constitution.*—An observation of fundamental importance is the direct conversion of cellulose into a crystalline furfural derivative under the action of the halogen hydracids. Empirically the reaction in the case of hydrobromic acid may be expressed by the equation:

$$C_6H_{10}O_5 + HBr - 3H_2O = C_6H_5O_2Br$$

the product being a brom-methyl furfural. The condensation takes place readily at 100° C. in presence of anhydrous ether. A particular point of interest arises in regard to the generalisation of the reaction as one specially characteristic of the ketoses, *i.e.* the keto-hexoses. The conversion of the typical levulose is represented as follows:

```
         ┌───O───┐                          ┌─H H─┐
    │    H    H    │                        │     │
 OH.C.C.  C    C  C  CH₂OH  =  OC.C : C.C : C.CH₂Br.
   H₂O   H OH  OH H                H └──O──┘
```

(H. J. H. Fenton. Chem. Soc. J., 1901, 361.)

The yields of the ω-brom-methyl furfural from various forms of cellulose were found to be high (33 p.ct.), higher indeed than from levulose. The reaction is therefore a main reaction, and shows that cellulose under these conditions breaks down, at least in large part, to ketohexose units. By these investigations therefore the polyaldose view of the constitution of cellulose is directly called in question. We have found on other grounds that a ketonic formula is to be preferred (1st ed. p. 77), the fifth O atom having ketonic rather than aldehydic function. This is consistent either with an open chain or closed ring formula for the assumed C_6 unit. There are general grounds of preference for the latter. But this is a matter of speculation and hypothesis.

A point to be noted in connection with Fenton's researches is that the normal celluloses (of the cotton group) give higher yields of the furfural derivative than the cereal celluloses (group C), which on the other hand are characterised by high yields of furfural under the action of aqueous condensing acids. This decomposition is referred by many chemists to the presence in the cereal celluloses of a pentose anhydride. In view of these later facts the explanation, which is on other grounds doubtful, becomes unnecessary. It is clear that the transition from the normal chain to the C_4O ring is equally characteristic of hexose as of pentose units, and the assumption that 'furfural yielding' is equivalent to 'pentose' or 'pentosane' carbohydrate, falls away.

A second point to be noted arises in connection with the exhaustive study of the action of ethereal hydrobromic acid on the celluloses. In a succession of treatments with the acid, diminishing yields of the brom-methyl furfural are obtained, and the final residue has the composition and character of the humic or ulmic series of complex derivatives described on p. 240. M. Gostling. Proc. Chem. Soc. 18, 250).

It is probable from later investigations of our own that pyrone groups are formed as an alternative or complementary course of condensation of the carbohydrates, and are represented in these complex products.

On the broad and general question of the actual constitution of cellulose there is as yet but little positive evidence. It is a question of proximate arrangement and configuration of ultimate constituent groups which we assume to be of C_n dimensions, and to be represented by the ordinary molecular formulæ. But we have no conception of a molecule of cellulose, and no data as to its dimensions. We have positive evidence as to a reacting unit, but of variable dimensions, and the more definite synthetical reactions of cellulose are expressed in terms of these units. But in these reactions the factor of mass, as distinct from relative molecular mass, has to be taken into consideration ; and, to cite a particular case, the recent elaborate investigations of W. Will on the nitration of cellulose, and the decompositions of the nitrates by heat, lead to the conclusion that in both directions there are no breaks of continuity corresponding with definite reacting units of relatively small dimensions. (*Infra*, p. 317.)

This problem of the relation of molecule to mass necessarily also arises in regard to the structural peculiarities of cellulose. The conversion of cellulose into films, threads, and generally into solids of continuous dimensions, has shown that the mechanical properties of these solids are a direct function of the molecular state of the parent substances, whether celluloses or cellulose derivatives. Thus the hydrocelluloses (p. 54) are formed from the fibrous celluloses at the expense of their tenacity : similarly, when converted through soluble derivatives into continuous solids, these are brittle and of low tenacity. The normal acetates give tough films ; but if the acetylation is carried to the point that *chemical* disintegration begins, as evidenced by the presence in the product of reactive CO groups, the product gives brittle films.

These considerations may be borne in mind in regard to the future investigation of the problem. But the problem is without present promise of solution, and it must be admitted that we have no criterion of the kind or degree of association of the molecular units in the complex aggregates of the cellulose group.

Esters.—On the general subject of the nitric esters of the carbohydrates Will and Lenze have made investigations leading to the conclusions that whereas the *aldoses* are fully esterified, the hexoses giving pentanitrates and the pentoses tetranitrates, the *ketoses* with n.OH groups yield nitrates containing $\overline{n-2}$. $O.NO_2$ groups as a maximum, the two remaining OH groups passing into the anhydride form. These nitrates of the ketose-anhydrides are distinguished by much greater stability. (Berl. Ber., 1898, 68.)

The authors have investigated the reaction of formation of these nitric esters under the usual conditions of treatment of the celluloses with a mixture of nitric and sulphuric acid, and conclude from their experiments that the latter acid reacts also with the cellulose hydroxyls. The fixation of SO_4H residues in some quantity is proved by analysis of the products formed under certain conditions ; and the fact has to be taken into account under all conditions of treatment, especially in regard to the very important question of 'stability,' and the uses of these products as explosives. (Cross, Bevan, and Jenks. Berl. Ber. 34, 2496.)

The highest derivative in this series of esters being the trinitrate—on the C_6 formula—the fact is shown to be consistent with the presence of 4.OH groups in the cellulose unit, which now must be taken as finally established by the general recognition of the highest acetate as a tetracetate, and as a true cellulose derivative. A higher degree of acetylation implies a hydrolysis of the cellulose, which is confirmed by a study of the properties of such derivatives. These conclusions have been verified and extended by the later investigations of Z. H. Skraup of the acetylation of starch and cellulose. (Berl. Ber. 1899, 2413.) In regard to the lower limits of acetylation it is stated in this volume (p. 35, 1st ed.) that the normal celluloses do not react with acetic anhydride at its boiling temperature. Investigations by the authors have shown that this statement, current in the text-books, is erroneous; a mono-acetate (C_n) is formed under these conditions. This product is insoluble in all the solvents of the cellulose esters, and moreover resists the action of cuprammonium solutions.

The authors have further investigated the benzoates of cellulose, and the conditions of their formation by interaction of cellulose and benzoyl chloride in presence of alkalis. From these esters mixed esters have been obtained by the action of nitrating acid. The benzoyl residues are converted into nitrobenzoyl, and further reaction ensues with the residual OH groups of the cellulose.

The following conclusions appear to be justified : the highest benzoate is the dibenzoate, or on the C_{12} unit the tetrabenzoate. Taking 8.OH groups as the maximum in this unit, five only react in these mixed esters, as compared with six as a maximum in the simple nitric esters. (See 'Researches on Cellulose,' pp. 34-40.)

From points of view other than the purely theoretical, various and important investigations of cellulose esters have been published in recent years.

Lunge and Bebie have carried out an elaborate enquiry into the constants of nitration of the normal cellulose ; chiefly concerning the yields and composition of the nitrates under definite variations of the more important chemical and physical conditions of the reaction. The results constitute the most extensive series of numerical records hitherto published, for which the original papers must be consulted. (Ztschr. Angew. Chem. 1901, 483. See also O. Guttmann, Chem. Ztschr. I. No. 12.)

Appendix II

The authors with A. Luck have also investigated the actions of diluted solvents upon the fibrous nitrates, under the action of which they are converted into dense structureless forms with elimination of the products causing instability. The process is the basis of technical developments based upon the more perfect control of the process of gelatinisation or 'colloidisation,' an essential condition of the use of these products as restrained or progressive explosion. (A. Luck and C. F. Cross. J. Soc. Chem. Ind., 1900.)

The most important event in connection with the scientific and technical development of this subject has been the foundation in Germany of the Research Institution of Neu Babelsberg, Berlin. (Central Stelle für Wissenschaftlich-technische Untersuchungen.) This institution, mainly devoted to the technology of nitrocellulose explosives, has published two brochures on the question of the stability of the cellulose nitrates. Full abstracts of these communications will be found in the Journal Soc. Chem. Ind. 1901, 609, 617; 1902, 1470-1. We can only notice here the main result of the elaborate investigations of Prof. Will in its bearing on the scientific side of the subject. It has been established that the normal stable nitrates when heated at high temperatures in an atmosphere of dry carbonic anhydride are continuously decomposed with a regular disengagement of nitric oxide, the decomposition taking place according to the typical equation:

$$C_{12}H_{15}(NO_2)_5O_{10} = C_{10}H_3NO_8 + 4NO + 6H_2O + 2CO$$

and reaching the limit represented by the formation of the end-product in question. The points to be noted in the composition of this product are the retention of one-fifth the original nitrogen and the loss of 1-C atom for each C_6 unit. Until the constitution of this empirical residue has been elucidated we cannot go beyond the statistical relationships established. The prominent general feature of the decomposition or dissociation is its regularity, *i.e.* continuity, upon which the 'stability' tests are based. It suggests a similar continuity in the original ester reactions.

We may briefly note here the publication of a book under the title 'Smokeless Powder, Nitrocellulose, and Theory of the Cellulose Molecule,' by J. B. Bernadou: New York, 1901 (London: Chapman & Hall, Ltd.). This work contains, in addition to the author's interesting speculations and records of experimental work, a résumé of important recent investigations of Vieille and Mendeljeff.

Cellulose Sulpho-carbonates (Viscose).—The authors have published an account of later researches into the nature and constitution of this series of compounds. The main point established is that the affinity of the cellulose xanthogenic acid is considerably higher than that of the fatty acids, and generally higher than that of the monocarboxylic acids. Consequently the solutions of the crude compound may be treated *e.g.* with acetic acid in excess without decomposing the alkali salts of the cellulose sulphocarbonic acid. The acetic acid, on the other hand, entirely decomposes the by-products of the original reaction and reactions of spontaneous decomposition. By this means the isolation of pure compounds of this series is much facilitated, the separation from sodium acetate on addition of alcohol being satisfactorily sharp.

The following stages in the process of reverse decomposition have been established: The general formula $CS\begin{subarray}{l}OX\\SNa\end{subarray}$ having been verified with satisfactory precision, and X being the cellulose residue of various dimensions, it is found that when freshly prepared X lies between C_6 and C_{12}, and the compound is not precipitated by dehydrating agents: as X approaches C_{12} the xanthate is precipitated by alcohol, and readily redissolves in water: the C_{24} xanthate is precipitated by smaller proportions of dehydrating agents from *alkaline* solutions, and is entirely precipitated by acetic acid; in other words, is insoluble in water. The cellulose when reaggregated to these dimensions is not soluble as a sodium xanthate, but requires the further combination of its OH groups with the alkaline hydrate to produce a soluble compound. These stages are well defined, and by their general recurrence in the course of investigations to the apparent exclusion of intermediate stages, it is suggested, though it cannot be finally affirmed, that the decomposition as it actually occurs in the solution takes place in the later stages by units of C_{12} dimensions.

The analysis of viscose solutions is obviously much simplified by these observations. By volumetric estimation, using successively normal acetic and hydrochloric acids, the alkali combined with the cellulose is determined, and the number can be confirmed by titration with a standard iodine solution. (Berl. Ber., 1901, 34, 1513-20.)

Ligno-celluloses.—The authors have shown that the colour reactions of the lignocelluloses with phenols are not characteristic of the lignone complex as such, but are due to break-down products—in all probability to hydroxyfurfurals. These bodies have been prepared by the interaction of furfural and hydrogen peroxide in presence of iron salts: they give reactions with phloroglucinol and resorcinol, identical with those of the lignocellu'oses in their natural state. (Cross, Bevan, and Briggs. Berl. Ber. 33, 2132.) These reactive constituents of the natural lignocelluloses are easily removed by treatment with oxidants in regulated small proportions, the lignocellulose undergoing only small losses of weight, and retaining its essential chemical characteristics unchanged.

Further studies of the lignone complex in the case of the jute fibre have somewhat modified the conclusions set forth in the first edition, and the text has been accordingly rewritten in those portions.

The furfural-yielding constituent is more probably a cellulose or an anhydride, and appears with the cellulose complex when isolated by the chlorination process. The lignone is thus to be considered as distinct from this β-cellulose and from the hydroxyfurfurals. These latter may be formed from the β-cellulose by processes of hydrolysis and condensation, and oxidation occurring 'naturally.' It is certain that active oxygen is always present on the surface of the lignocelluloses, indicating a slow and progressive auto-oxidation of the fibre substance. The observations and ingenious investigations of W. J. Russell (Nature, vol. 65, p. 200) have emphasised these phenomena by showing that they are associated with 'emanations' which act upon sensitive photographic surfaces, and produce an image of the objects. Russell considers that the evidence so far accumulated points to these emanations being hydrogen peroxide.

The problem of the constitution of the characteristic lignine complex is so far simplified. Its most important constituent groups are: (1) the benzenoid group, combining directly with chlorine; and (2) a group or groups of approximate formula $C_{2m} H_{2m} O_m$, which break down by gentle oxidation and hydrolysis finally to acetic acid as a main product, with probable formation of ketonic acids of low molecular weight as intermediate stages. The complex contains a minimum proportion of hydroxyl groups and of methoxyl groups.

Some further light has been thrown on the relationships of

cellulose to lignone groups in the lignocellulose complex, by later investigations of certain lignocellulose esters.

The benzoate prepared by treating with benzoyl chloride in presence of alkali is a monobenzoate, calculated to the simplest empirical formula $C_{12}H_{18}O_9$. The benzoyl group enters the cellulose residue ; the lignone is unaffected, and when removed by the ordinary treatment a cellulose benzoate is left as the end product.

On boiling the lignocellulose with acetic anhydride, an acetate is formed, which analyses as a diacetate of the empirical unit $C_{12}H_{18}O_9$. The complete statistics, however, appear to show that the ester reaction is attended by internal dehydration through interaction of other groups of the complex. The lignone group, however, retains its general characteristics, and may be removed by similar treatment as the original, and the cellulose is separated in the form of a diacetate (C_{12}).

The benzoate (supra) also reacts with acetic anhydride, and the proportion of acetyl groups entering is not affected by the presence of the benzoyl group.

These ester reactions taking place in the cellulose group, it is further established that the lignone complex contains no OH groups reactive under these conditions, and also that there are no free aldehydic groups.

These reactions are of use in the investigation of the ultimate constitutional problems which continue to engage the attention of the authors, and to which it is hoped other chemists will be attracted by the publication of these evidences of more definite progress.

INDEX OF AUTHORS

ABEL, 44
Armstrong, 245

BAEYER, 245
Bary, de, 231
Bebie, 316
Béchamp, 46
Beilstein, 259
Benedikt and Bamberger, 188, 232
Bernadou, 317
Berthelot, 87
Briggs, 319
Brown, A. J., 72
Brown, Horace, 67
Brown and Morris, 65, 257
Brunner, 167

CALVERT, Crace, 21
Chalmot, de, 181, 185
Chardonnet, de, 45
Chevandier, 174
Chodnew, 216
Collie, 62, 149
Cross and Bevan, 7, 61, 70, 79, 80, 83, 113, 124, 131, 137, 138, 142, 152, 164, 208, 232, 240, 244, 247, 259, 263
Cross, Bevan, and Briggs, 319
Cross, Bevan, and Jenks, 316
Cross, Bevan, and King, 243

Cross and Witt, 148
Crum, W., 24

DEMEL, 240
Döpping, 226
Durin, E., 72
Du Vivier, 45

ERDMANN, 11, 161, 197

FENTON, H. J. H., 313
Fischer, E., 262
Fischer and Schmidmer, 18
Flechsig, 49
Flint and Tollens, 99, 265
Flückiger, 228
Franchimont, 61, 87
Francis, 254
Frank, 224
Frémy, 90, 91, 173, 176, 216, 229

GANS and Tollens, 221
Gilson, 12
Girard, 21
Godeffroy, R., 219
Goodale, 237, 243
Goppelsroeder, 18
Gostling, M., 314
Gottlieb, 175
Green, Cross, and Bevan, 298

Griffin and Little, 288
Guignet, 53
Guttmann, O., 316

HALLIBURTON, 87
Hantzsch and Schniter, 137
Hawes, G. W., 175
Hime and Noad, 10
Hodges, 80, 232
Hoehnel, 227
Hofmann, A. W., 79
Hofmeister, 237
Hönig and Schubert, 48, 225
Hoppe-Seyler, 66

KABSCH, 173
Karolyi, 44
Kirchner and Tollens, 221
Knecht, 55
Koechlin, C., 21
Kolb, 218
Krauch and V. d. Becke, 166
Kraus, 153
Kugler, 227
Kuhlmann, 46

LANGE, 22, 141, 214, 240
Lehner, 45
Lenze, 315
Lindsey and Tollens, 49, 83, 198
Lintner and Düll, 257

Lloyd, 18
Löwig and Kölliker, 87
Luca, de, 46
Luck, A., 317
Lunge, G., 316

MACNAB and Ristori, 44
Mann and Tollens, 184
Maurey, 46
Meissner and Sheppard, 152
Mendeljeff, 318
Mercer, 24
Meyer, V., 163
Miller, W. A., 239
Mitscherlich, 226
Mühlhäuser, 132
Müller, Hugo, 5, 79, 110, 175, 214, 219
Müntz, 168

NASTJUKOW, 61
Nölting and Rosenstiehl, 61

O'SULLIVAN, 257

PARNELL, 24
Payen, 211
Pears, A., 111

Pelouze, 46
Poumarède and Figuier, 187
Prudhomme, 11

RAMSAY and Chorley, 68, 154, 204
Reichardt, 216
Rosenfeld, 13
Russell, W. J., 319

SACHS, 73, 237
Sachsse, 172, 221
Schaefer, 87
Scheibler and Mittelmeier, 258
Schleiden, 224
Schlichter, 272
Schmidt, 61, 87, 225
Schmitz, 237
Schulze and Tollens, 163, 260
Schunk, 290
Schuppe, 177
Scoffern and Wright, 13
Sestini, 137, 240
Skraup, Z. H., 316
Smith, 83, 164, 259
Spon, 79, 243
Stein, 271
Stern, 49

Stutzer, 152

TAUSS, 22
Thomsen, 187
Thorn, 213
Tollens, 101, 181, 259, 261
Tollens and others, 261

URBAIN, 173

VÉTILLART, 243
Vieille, 41, 318
Vortmann, 265

WATTS, 291
Weber, 19, 131
Webster, 113
Weiske, 152
Wheeler and Tollens, 187, 212
Wiesner, 243
Will, 47, 315, 317
Wissenburgh, 228
Witt, O. N., 79
Witz, 61, 297
Wurster, 174

ZEISEL, 106, 189, 265

INDEX OF SUBJECTS

ACETATES of cellulose, 34, 252, 316
Acetic acid, formation from cellulose, 255
— anhydride, action upon cellulose, 35, 316; upon regenerated cellulose, 37
— condensation, 193
— residue in woods, 191; product of simple hydrolysis of lignocelluloses, 192; characteristic feature of lignification, 192, 319
Adipocelluloses, 90, 225, 226; proximate analysis, 227; general methods of investigation, 267
Aloe fibres, 220
Amylobacterium, 66
Amyloid, 53, 224
Aniline dyes, action on jute fibre, 115
— salts, action on jute fibre, 115
Arabic acid, 216
Arabinose, 86, 216
Arabinosic acid, 216
Ascidia, 87

BACTERIUM xylinum, 73
Ballistite, 44, 309
Bamboo stems, 220
'Belfast Linen Bleach,' 80
Benzoates of cellulose, 32, 251, 316
Blasting gelatin, 309
Bleaching, isolation of cellulose from raw fibres, 244, 255; linen yarn, 286; jute cuttings, 287; cotton, 288; 'market bleach,' 'madder bleach,' 291; linen, 292

'Brewers' grains,' composition of, 163, 260; method of examination, 260
Brom-methyl furfural, 313
Butyric fermentations, 234

CARBOHYDRATES, 2; general methods for identification, 261; nitration, 315
Carragheen mucilage, 225
Celluloid, 44, 308
Cellulose, 1; empirical composition, 3; hydrates, 4; their reaction with iodine, 7; of green fodder plants, 7; solutions of, 8; in zinc chloride, 8; in zinc chloride and HCl, 9; in ammoniacal cupric oxide, 9; in ammoniacal cuprous oxide, 13, 246; threads or filaments in electric lamp, 8; crystallised, 12; theory of action of solvents, 14; qualitative reactions and identification, 14; compounds of, 15; with dilute alkalis and acids, 16; with colouring matter, 19; capillary phenomena, 18; action of alkaline solutions at high temperatures, 22; action of concentrated alkaline solutions, 23; thiocarbonates, 25, 318; their spontaneous decomposition, 26; their coagulation by heat, 27; quantitative regeneration of cellulose from solutions of thiocarbonate, 28;

purification by alcohol and by brine, 248 ; uses in microscopic work, 249 ; theoretical notes, 249 ; regenerated cellulose from thiocarbonate, 29; reaction with acetic anhydride, 37 ; theoretical view of thiocarbonate reaction, 29, 316. Reacting unit, 31 ; benzoates, 32, 316; soluble alkali, 33 ; acetates, 34 ; interactions with acetic anhydride, 35 ; and acetic anhydride in presence of zinc chloride, 36 ; in presence of iodine, 36 ; nitrates or nitrocelluloses, 38 ; their general properties, 39 ; approximate composition (table), 42 ; thermal constants, 42 ; heat of combustion, 43 ; products of combustion, 43 ; industrial uses, 44, 307 ; gradual decompositions, 46. Action of sulphuric acid, 48 ; transformation to a sugar, 49 ; composition of body produced by dissolving in HSO_4, 49. Decompositions of, 52 ; by non-oxidising acids, 53 ; practical application, separation of cotton from wool fabric, 55 ; by oxidants, 56 ; in acid solutions, 56 ; in alkaline solutions, 60 ; resolution by ferments, 63 ; resolution constituting 'decay,' 66 ; by condensation of carbon nuclei, 66 ; feeding or nutritive value of, 67 ; destructive distillation of fibrous, of regenerated from thiocarbonate, 68 ; tables, 69 ; constitution of, reactions throwing light upon it, 75 ; theoretical notes on, 257 ; three subdivisions in group, 78 ; purification in laboratory, 79 ; 'cellular,' 82, 85 ; from woods and lignified tissues, 83 ; elementary composition, 83 ; yield of furfural, 83 ; from cereal straws, esparto, 84 ; their ultimate composition, 84 ; yield of furfural and reactions, 84 ; results from solution as thiocarbonate, 85 ; regenerated from straw and esparto cellulose thiocarbonate, 85 ; pseudo- or hemi-, 87 ; a constituent of protozoa, 87 ; compound, 89 ; adipo- and cuto-, 90 ; pecto- and muco-, 90 ; Frémy's classification, 90 ; para- and meta-, 90 ; ligno-, 91, 92 ; (see Jute) α and β, from jute fibre, 93 ; general view of the group, 235 ; processes of decay and destruction (tables), 239 ; morphology, 243 ; technology, principles of, 273 ; preparation of fibres from raw material, 276 ; flax, and jute, 275 ; spinning, 279 ; bleaching, 284 ; of jute, 285 ; linen yarn, 286 ; jute cuttings, 287; cotton, 288 ; constitution of, 313

Cell-wall, differentation of substances composing, 86
Cerin, 228
Ceryl alcohol, 80
China grass, 79, 220, 278
Chloroplasts, 73
Coal, 66, 238
Collodion varnishes, 44 ; films, 44
Colloidal cellulose, 53
Combustion, rapid method, 245
Condition, water of, 5
Cordite, 44, 309
Cork, 225, 226
'Crude fibre,' Weende method of estimation, 165
Cutin, 228
Cutocelluloses, 90. See Adipocelluloses
Cutose, 90, 229, 230

DECACRYLIC acid, 227
Dehydration, 245
Dextrose, 64, 74, 86, 222, 261
Diastase, 71 ; secretion by flowering plants, 74
Diazotype process, 298
Drupose, 162
Dye woods, 204
Dyeing processes, 294
Dynamite, 309

ELECTRIC lamp, 8
Enzyme (cyto-hydrolyst), 65; in digestive tract of herbivora, 67, 216
Esparto, 84, 220
Eulysin, 227
Explosives, 44, 308, 317

FERMENT, acetic, forming cellulose, 72; hydrolyses of cellulose, 255
Ferric ferricyanide, action on jute, 115; theory of dyeing, 124; behaviour with gelatin, 129
Fibres, raw, investigation of, 269; fibre constants, 300; numerical expression, 301
Films, 4, 44
Filter paper, 3
Finishing processes, 6
Flax, 79, 80; retting and scutching, 217; cortical tissue, 217; flax fibre proper a pectocellulose, methods of isolation in laboratory and in practice, 218; flax cellulose, 219
Food in relation to work, Müntz's researches, 168; tables, 169, 170, 171
Frémy's classification, 173
Furfural, product of acid hydrolysis of oxycelluloses, 82; reagent for obtaining, 82; furfural-yielding complex, 98; estimation of, 99; oxidation of, 319. *See also* Jute

GALACTOSE, 86, 199, 216, 222, 262
Glycerol, 228
Glycodrupose, 161
Glycolignose, 198
Glycuronic acid, 184
Green fodder-plants, 7; investigation of, 270
Grundsubstanz, 167
Gum-arabic, 216
Gun-cotton, heat and products of, combustion, 43. *See* Cellulose nitrates

HACKLER'S dust, 234
Hackling, 80, 234
Hemi-celluloses, 87

Hemp, 79
Hexoses, identification, 261
Hippuric acid, 192
Humus, 238, 239, 314
Hydracellulose, 54
Hydration, 245
Hydrocellulose, 54
Hydroxyfurfural, 317
Hydroxypyruvic acid, 47

JUTE, 91; composition of fibre, 92; furfural-yielding complex, 92; cellulose isolated not homogeneous, 93; quantitative estimation of cellulose constituents, 94; by chlorination, 94; by bromination, 95; by treatment with nitric acid and potassium chlorate, 96; with dilute nitric acid, 97; by sulphite and bisulphite process, 97; estimation of furfural-yielding complex, 99; of keto R. hexene constituent, 101; estimation of constants of chlorination, 102; empirical formula, 102; determination of HCl in reaction, 104; control observations, 104; estimation of secondary constituents (methoxyl) by standard method of Zeisel, 106; $CO.CH_2$ residue, 107; systematic account of fibre; 'butts' or 'cuttings,' 109, 287; sp.gr., 110; analysis of various specimens (table), 110; composition, 111; empirical formula, 111; artificial cultivation by A. Pears, 111; analysis of cultivated fibre (table), 112; lignocellulose hydrates, 113; solutions of lignocellulose, 114, 263; qualitative reactions and identification, 115, 262; action of aniline salts and coal-tar dyes, 115, 262; action of phloroglucinol, iodine, chlorine, ferric chloride, ferric ferricyanide, 115, 262, 266; chromic acid, potassium permanganate, 116; compounds of jute cellulose, with acids and alkalis, from dilute solutions, 116, 263; from concentrated solutions of alkaline hydrates, mercerisation, 120; thiocarbonate reaction, 121; compounds with metallic salts, reaction with ferric ferricyanide, and theory of dyeing, 124; compounds with negative radicals, benzoates, 131; acetates, nitrates, 132; lignocellulose under nitration behaves as a homogeneous body, 134; compounds with the halogens, chlorine, 134; bromine, 137; iodine, 138, 263; resolution into constituent groups, 139; by hydrochloric, hydriodic, sulphuric acid, 140; nitric, dilute, in presence of urea, 141, 264; by alkalis, 141; by acid oxidants, chromic acid, 142, 264; chromic and sulphuric, 144; strong nitric, 145; joint action of oxides of nitrogen and chlorine, 147; by alkaline oxidants, potassium, permanganates, 147; hypochlorites, 148; hypobromites, 149; interaction with sulphites and bisulphites, 149, 265; animal digestion, 151; spontaneous decomposition, 152; destructive distillation, 153; general conclusions as to composition and constitution of lignocellulose, 155; lignocellulose considered as a whole, 157, 319; ultimate analysis, 266

KETO R. hexene constituent of lignocelluloses, 101. *See* Jute
Kieselguhr, 309

LEUCOGALLOL, 102
Levulinic acid, 199
Levulose, 74, 262; condensation of, 313
Lichenin, 214
Lignification, 92
Lignin, 94
Lignite, 66, 238
Lignocellulose, 91, 92; esters, 131; of cereals, composition of brewers'
grains, 163; straws, how differentiated from typical lignocellulose, 164. *See* Jute
Lignone, 94, 319, 320

MAIROGALLOL, 102
Maltose, 74
Mannose, 86; hydrazone, 199; identification, 262
Meals, analysis of, 167; rice meal, 167
Mechanical wood-pulp, estimation in paper, 174
Mercerisation, 23, 120
Metapectic acid, identical with arabic acid, 216
Metapectin, 216
Methoxyl determination in woods, 106, 188, 265
Mitscherlich process, 198
Mucic acid, 199
Mucocelluloses, 90. *See* Pectocelluloses
Musa, 220

NITRIC acid, toughening action on papers, 254
Nitrocelluloses, or cellulose nitrates, 38, 309, 315, 316
Nitroglycerin, 309

OIL-WAX complex in flax, 80
Oleocutic acid, 230
Ophrydium versatile, 87
Oxycellulose, 56, 82; preparation and diagnosis, 254; identification, 262
Osazone, 222

PAPER, analysis of, 271; permanence a first desideratum in writing and printing paper, 304; disintegration of modern, 306
Paper-making fibres, 280
Parapectic acid, 216
Parapectin, 216
Parchmenting process, 253

Peat, 66, 238
Pectase, ferment enzyme, 216
Pectic acid, 216, 290
Pectin, 216
Pectocelluloses, 90; how distinguished from mucocelluloses, 215; general characteristics, 217; flax, 217 (*which see*); China grass or Ramie, nettle fibres, monocotyledonous fibre aggregates, 220; parenchymatous tissue of fruits, 221; mucilaginous constituents of plant tissues, quince mucilage, 221; salep mucilage, 223; amyloid, lichenin, 224; carragheen mucilage, 225; general methods of investigation, 267
Pectose, 90, 216
Pentaglucoses, 93, 262
Pentosans, 93, 185, 186
Pentoses, 86
Phloroglucinol, action on jute, 115, 192
Phormium, 220
Powders, smokeless, 309; sporting, 309
Printing processes, 294
Pseudocarbons, 70
Pseudocelluloses, 87
Pyrocatechol, from woods of Coniferæ, 198
Pyrocatechuic acid, 198
Pyroxylins, 39

QUINCE mucilage, 221

RAMIE, 220
Retting, 67, 80, 234, 277
Rhea. *See* China grass
Rot-steep, 67

SACCHARIC acid, 261
Salep mucilage, 223
Schultze's reagent, 173
Scutching, 80
Silk, Dr. Lehner's artificial, 45, 308
Skeletonising, process of, 66; a simple means of differentiation, 67

Spinning processes, 279
Stability of nitrocelluloses, 317
Stearocutic acid, 230
Straws, cereal, behaviour in thiocarbonate reaction, 164; wood gum in, 187
Suberin, 228, 231
Suberose, 228, 231
Sugar cane, 220
Sugars, 65; cane sugar first assimilated, and probable immediate mother substance of cellulose, 72

TEXTILES, analysis of, 271
Thiocarbonate of cellulose, 25, 318. *See also* Cellulose
Tissue-substance, first step in building-up of, 74
Tunicin, 87

VARNISHES, collodion, 44
Vasculose, 90
Viscose, 25, 247, 318

WEENDE method, 165
'Willesden' goods, 13
Wood-gum, 187
Woods, 91; structural elements, 172; Frémy's classification, 173; general property to form hydrogen peroxide, and estimation of mechanical wood-pulp in papers, 174; empirical composition, 174; tables, 75; proximate analysis table, 175; resolution into cellulose and non-cellulose, 177; discussion of Sachsse's view that lignocelluloses are the products of metabolism of cellulose, 178; estimation of furfural (table), 182; de Chalmot on life-history of woods, 182; wood gum, 184, 187; in cereal straws, 187; analyses, 188; methoxyl determinations, 188, 189; acetic residue, 191; destructive distillation, 192; chlorination of wood lignocelluloses, dicotyledo-

nous, 194; coniferous (table), 195; synthetical reactions, nitration, 196; chemistry of woods of Coniferæ, 197; investigation of sulphite pulp process, 198; empirical formulæ with methoxyl determinations, 201; yields of pulp, 202; destructive distillation, 204; tables, 205, 256; Ramsey and Chorley's tables, 207; disintegration by reagents, proximate resolutions, 208; table, 209; sulphurous acid, bisulphites, neutral sodium sulphite, 210; alkaline processes, acid processes, 211; ultimate resolutions, extreme action of alkaline hydrates, 213; chromic, in presence of sulphuric, acid, 214, 260

XANTHATES of cellulose, 26, 247, 318. *See* Cellulose thiocarbonates
Xylan, in wood gum, 184
Xylonite, 44, 308
Xylose, 86

'ZEISRL,' standard method of estimating methoxyl, 106, 189, 205

www.ingramcontent.com/pod-product-compliance
Lightning Source LLC
Chambersburg PA
CBHW031424230426
43668CB00007B/419